PETER CAIRNS   FLORIAN MÖLLERS   STAFFAN WIDSTRAND   BRIDGET WIJNBERG

WILD
WONDERS
OF EUROPE

ABRAMS, NEW YORK

Contents

Joy of the Wild

The sudden excitement as a fox darts across the road right in front of you, a blackbird's sweet song drifting through your window from a nearby rooftop; a skein of geese passing high above the city where you work, a flower-filled woodland lighting up your morning as you cycle to school, the croaking of frogs announcing spring, the breathtaking skyline of snowcapped mountains on the distant horizon. These are all wild places and wild beings; and when we meet them, however briefly, it tends to make most of us feel a little happier and a little more alive.

They are all part of an ancient web that is the very foundation upon which our own lives depend. Always have, always will.

They are part of Europe's Wild Wonders, and that is what this book is all about.

Meeting wild beings tends to make us feel good.

The garden hedgehog or robin, as well as the soaring eagle. They make us smile and even stop in amazement. They get us all excited and make us want to tell other people. Wild things stir emotions deep within us. We don't own them. We can't really control them. They root or roam wherever they like. It's healthy for our cluttered minds to know that there are still things out there, wild and free, that are not part of any city plan or corporate strategy—creatures or plants that simply do their own thing. It doesn't matter what age we are: meeting the wild brings joy to young and old alike.

It is the same with wild places. They symbolize freedom and enduring values like beauty, serenity,

calm, and dignity. They encapsulate feelings of hope and heritage, providing us with some much-needed peace of mind.

We all firmly believe that this "joy of the wild" is deeply rooted in our genes and that this is one of the cornerstones of human happiness.

That is why we want to celebrate the fact that Europe is still full of exciting wildlife and wild places.

And more so than most people think. We want to reveal the amazingly exotic, varied, and charming natural gems of the European continent. The best of the best, all in their splendour. We want to show that the Europeans have a shared, vibrant natural heritage, which most still need to learn much more about. This heritage is a vital part of European

**Loggerhead sea turtle
hatchling rally**
Caretta caretta
TURKEY/DALYAN DELTA
Since ancient times, sea turtles have
laid their eggs on most sandy beaches
around the Mediterranean. Now they are
almost completely gone, due to
centuries of over-harvesting their eggs
for food, bycatch during fishing; and
more recently, insensitive hotel and
housing construction on their breeding
beaches. Huge efforts have been made
to protect the final few egg-laying
beaches and to help more of the hatch-
lings survive. The most important turtle
beaches in Europe today are in southern
Turkey, and local World Wildlife Fund
staff have done an impressive job trying
to save them. Now things are beginning
to look better for several sea turtles.
But there are still appalling losses of
turtles to bycatch in high-sea fisheries.
Solvin Zankl

Tulip close-up
Tulipa doerfleri
GREECE/SPILI, CRETE
This rare tulip is endemic to Crete. It is
one of several wild tulip species that are
the origins of our common garden
tulips. The genetic variety found within
these wild species has made it possible
to create the myriad domestic tulips we
see today. The variety of life, or biodiversity, in action!
Peter Lilja

identity. Getting to know it means
getting to know ourselves.

As humans, we can only really
care about things we have strong
feelings for. But how are we
supposed to have feelings for
something we don't even know
exists? And how are we supposed to
vote for it? This is why we want to
share the unknown Wild Wonders of
Europe, the crown jewels of nature in
Europe, with you.

We also want to bring you some really good news.

Wildlife in Europe is coming back!
It's a fantastic success story. After
decades, even centuries of
humanity's wasteful practices,
Europe's wildlife is starting to return!

That is not to say that every-

thing is fine. We all know it isn't. We
hear about the problems day and
night, and we certainly need to work
hard to put many things right. But
the glass is not half empty, it is
actually half full. And it is
getting fuller.

We want to celebrate the reasons
why these wild animals, plants, and
places are still with us in spite of the
enormous pressure we have put on
them. That is because Europeans
have decided to do a few things a bit
differently. We have made a number
of nature-conservation decisions.
And, surprise, surprise, when we
treat our fellow species with a bit
more tolerance and respect, we are
immediately rewarded with their
return. This book is therefore a cele-
bration of nature conservation in

LEFT
Short-snouted sea horse
Hippocampus hippocampus
MALTA/GOZO

The Mediterranean Sea is one of the world's 25 biodiversity hotspots. The variety of species, both in the sea and on its many islands, is simply staggering. But centuries of over-exploitation means there are fewer and fewer individuals of each species now surviving, except in a scattering of marine reserves like Dwejra in Malta.

Solvin Zankl

UPPER RIGHT
European tree frog
Hyla arborea
BULGARIA/NIKOPOL, PLEVEN

The four species of tree frog in Europe can be found from the Mediterranean to southern Scandinavia. The tree frog is smaller than a matchbox, but with a very loud, ringing call that can be heard more than a kilometre (almost a mile) away.

Dietmar Nill

LOWER RIGHT
Fishermen taking up their catch
ALBANIA/LAKE PRESPA NATIONAL PARK

Lake Prespa produces lots of fish, enjoyed by birds and man alike. Historically, all fish-eating birds were heavily persecuted here; but now they are protected. As a result, dwarf cormorants, Dalmatian pelicans, grebes, storks, and several heron species are increasing in numbers and can be seen by the thousands once more. The lake is part of the Transboundary Prespa Park created in 2000, linking Albania, Greece, and Macedonia.

Anders Geidemark

PAGES 24–25
Osprey at work
Pandion haliaetus
FINLAND/KANGASALA, PIRKANMAA

The osprey is a true global traveller and is found across every continent except Antarctica. Europe's ospreys breed mainly in the north—in Finland, Sweden, and Russia—and travel south for the winter. This fish-eating specialist has made a comeback in recent years due to a reduction of toxins in its food, less persecution, and better protection.

Peter Cairns

PAGES 26–27
Baltic Sea beach
DENMARK/MØNS KLINT

The limited influx of salt water from the Atlantic, combined with fresh water from the many rivers that feed the Baltic Sea, creates a unique brackish habitat. Essential steps have been taken in recent years to reduce pollution and sewage reaching the Baltic, but nutrients leaking from farm and forest fertilizers still damage the sea's fragile chemistry.

Sandra Bartocha

Europe celebrating what has already been accomplished and highlighting some of what still needs to be done.

Finally, it is about a wonderful love affair!

We want to inspire millions of people to spend more time in the great outdoors: to go out into nature, to enjoy and explore it firsthand with family and friends. Nature really deserves to be loved; and like any love affair, it needs to be a hands-on experience!

To be able to show all this to the world, we created an initiative that turned into the biggest nature photography project ever carried out: The Wild Wonders of Europe. We sent out 68 of the continent's best nature and wildlife photographers on an epic quest: to try to reveal Europe's amazing natural treasures to the world.

This photographic dream team was sent on more than 125 assignments, to all of Europe's 48 countries. They are now back and this is some of what they found for us all to absorb, reflect upon, and enjoy. Please join us on an image odyssey through the most precious Wild Wonders of Europe.

Unseen, unexpected, unforgettable...

Peter Cairns, Florian Möllers,
Staffan Widstrand, and Bridget Wijnberg

Unseen

Many believe that Europe long ago lost all its nature under a blanket of concrete and highways. The good news is that it hasn't. Not yet, anyway. It is mostly just unseen. When it comes to nature, many Europeans today seem to know more about Africa or North America than about their own natural heritage. Few have ever heard of creatures such as the European bison, the wolverine, the basking shark, or the saiga antelope. Few are aware that tens of thousands of vultures breed in southern Europe. And how many know about wild places like Plitvice, Landmannalaugar, Tichá, Bagerova, Gredos, Padjelanta, or Durmitor? From the knife-edged ridges of the Caucasus mountains to the forests of Białowieża and the coastal wetlands of the Wadden Sea, the time has come to unveil some of these gems of nature to the world.

Through the winter mists of the Białowieża forest in Poland, the huge shadow of a bull wisent, or European bison, is barely visible between the centuries-old tree trunks. In the frigid air, the tendrils of its breath push and probe through the trees, its frosted fur glistening as the first rays of a weak winter sun puncture the blanket of dawn. This primeval scene is barely imaginable in a modern, crowded continent seemingly stripped of so much of its wildness. Yet this old-growth forest survives even today, and so do the bison. And they're coming back. At one stage there were only 13 of these animals in existence—all in zoos. But these magnificent beasts are now returning to centre stage where they belong. To Poland and Belarus, the Russian Caucasus, Ukraine, the Netherlands, Slovakia, Latvia, Lithuania, and soon to more countries, such as Germany. From a couple of handfuls of animals in zoos, there are now about 2,000 reclaiming their rightful place in the wild. A hard-won, major victory for conservation.

There were once vast herds of these impressive buffalo roaming all across Europe, together with mammoths, aurochs, red and roe deer, moose, and wild boars. These grazers and browsers shaped a mosaic of forests, wetlands, and open plains, and were hunted by lions, wolves, and the ancestors of our ancestors. The lions, mammoths, and aurochs are sadly gone forever; but the bison are still here. A bison that looks pretty much the same as that portrayed in the 20,000-year-old cave paintings in Lascaux in France or Altamira in Spain. A symbol of our past connecting us to our present. The survival of the wisent is a potent reminder that the wild can endure and even prosper if we just allow it to.

RIGHT AND PAGES 32–33
European bison
Bison bonasus
POLAND/BIAŁOWIEŻA FOREST NATIONAL PARK

This is Europe's largest herbivore, weighing in at over 1,000 kg (2,200 lbs.). Hunted as close to extinction as any animal can be, there are now around 2,000 European bison roaming in the wild and another 2,000 in captivity. The Białowieża forest, straddling the border between Poland and Belarus, is the area where the bison, or wisent, clung on and survived. Białowieża is one of the very few remaining European old-growth broadleaf forests, and is constantly threatened by logging interests. There are plans to expand the protected areas on the Polish side. Centuries ago, these forests were probably more open and parklike, due to an abundance of large grazers and browsers.
Stefano Unterthiner

"In Białowieża, I felt as if I had jumped into the past to discover what central European forests may have looked like a thousand years ago.

Stefano Unterthiner, Wild Wonders photographer, Italy

Slowly, Europeans are beginning to realise that our natural heritage is shared. It is older and deeper than all of our national cultures and traditions. It is the very foundation of our evolution as a species and as a society in this part of the world. In the not-so-distant past, our very survival depended on what nature could provide: olive trees; grape vines; cork oaks; pine trees; parsley, sage, rosemary, and thyme; raspberries; lingonberries; blueberries; cloudberries; honey bees; deer; wild boars; bison and moose; seals; partridges; cod; herring; tuna; salmon; walnut trees; hazel bushes; chestnut trees; and boletus mushrooms. This natural harvest sustained and nurtured us. It sheltered us and kept us warm, and although it is tempting to think that we no longer need nature in our modern lives—where much of it goes unseen—we've not really changed that much ourselves. Perhaps it is at our peril that we turn our backs on the services that wildness provides.

All these species and habitats belong to this shared heritage. The distribution of plant and animal species is not at all related to national borders. Instead, it follows habitats, climate, and soil conditions right across the boundaries set by man. Bird species breed in one country, spend autumn and spring in another, and winter in a third or fourth. The common crane, for example, breeds in Scandinavia, Finland, Germany, and Russia. It migrates in large numbers through Germany and France and winters in Extremadura, in Spain. Another example is the knot, a small wading bird that breeds in the hundreds of thousands in Greenland and the Russian Arctic but spends part of the year on the wetland coasts of western Europe and western and southern Africa. Or the impressive raptor migration points in Tarifa, Spain; the Messina Strait, Italy; or Istanbul, Turkey; where the majority of all migrating European birds of prey pass during a few intense weeks each spring and autumn. It is all connected: if quail and turtle doves are hunted too heavily during their migration through southern Europe, they disappear from their northern breeding areas. If the arctic goose species are too heavily hunted in the northern countries, there are much fewer of them wintering in the southwest and southeast.

PAGES 34–35
Veliki Prstvaci waterfalls
CROATIA/PLITVIČKA LAKES NATIONAL PARK
The turquoise-coloured lakes and pools of Plitvička are fed by springs and runoff from the mountains, pouring out over the limestone cliffs in spectacular waterfalls. This is one of Europe's true gems of nature, previously almost unknown outside the Balkans. During summer, it is now visited by some 10,000 people daily.
Maurizio Biancarelli

RIGHT
Great bustard display
Otis tarda
SPAIN/LA SERENA, EXTREMADURA
With males weighing up to 16 kg (35 lbs.), the great bustard is one of Europe's true heavyweights. In spring the males perform an impressive display to attract the females, turning themselves almost inside-out into a ball of down and feathers. Previously a common bird in dry and steppe areas across the whole of Europe, the great bustard disappeared from most of its range because of hunting and habitat loss. It is now slowly coming back. The largest numbers are found in Spain, Portugal, Russia, Ukraine, Slovakia, and Hungary; and it is being reintroduced to Germany and the United Kingdom. It is the national bird of Hungary.
Staffan Widstrand

PAGES 38–39
Brent geese in Hallig Hooge
Branta hrota
GERMANY/WADDEN SEA NATIONAL PARK, SCHLESWIG-HOLSTEIN
Migrant visitors from the Russian Arctic tundra, Brent geese gather in large numbers together with other wildfowl in the tidal wetlands of the Wadden Sea, which stretches 400 km (250 mi.) from the Netherlands to Denmark. This is 10,000 sq. km (3,900 sq. mi.) of almost primordial landscape. It is one of the most important areas for water birds on earth, with some 10 to 12 million birds stopping over and another million breeding each year. On the German side, most of the area is included in the large Wadden Sea National Park. It has been a Unesco World Heritage Area since 2009.
László Novák

It is all connected. There are alpine plant species that are the same, or very closely related, across the Scandinavian mountains, the Alps, the Pyrenees, the Carpathians, the Balkans, and the Caucasus. The lowland species are also connected to each other, and so are the coastal species, the Mediterranean Basin dry-ground species, the central European forest species, the steppe species, and the boreal-forest species.

Many of our most cherished garden flowers and pot plants are developed from wild, original species in Europe: tulips, peonies, irises, daffodils, fritillaries, primulas, roses, lilacs, geraniums, and ivies.

This ancient natural heritage is really one of the fundamental, but rarely considered, answers to the question, What is Europe? Maybe it is actually even the first and most definitive answer. The natural environment that existed here set the conditions that shaped all the cultures that developed in this part of the world.

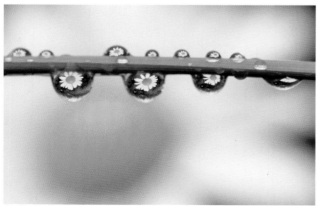

Meadow saffron
Colchicum autumnale
HUNGARY/DUNA-DRÁVA NATIONAL PARK

One of the few flower species in Europe that blooms in the autumn, the meadow saffron prefers natural grass-lands, a type of landscape that is today disappearing over much of the continent.

Florian Möllers

Mountain daisy
Leucanthemum adustum
LIECHTENSTEIN/MALBUN

Loves me, loves me not, loves me … The daisy is one of our most popular garden plants, and the mountain daisy is one of the original wild species.

Edwin Giesbers

The blue hour at Los Hervideros
SPAIN/LANZAROTE, CANARY ISLANDS

Lanzarote is the driest of the Canary Islands and has become a mecca for beach holidays. Yet remarkably large areas of Lanzarote are protected as reserves and national parks. The Canary Islands receive more than 11 million tourists each year and still protected Natura 2000 sites cover some 30 per cent of the islands' land surface.

Iñaki Relanzón

Mountain cornflower
Centaurea montana
LIECHTENSTEIN/ MALBUN

A flowering alpine mountain meadow is one of the most spectacular displays of nature's colour. Many of the species here depend on the regular hay harvest, and they quickly disappear if the area is chemically fertilized or if it turns into scrub or forest. These meadows are also very important to a great number of insects, especially many butterfly species.

Edwin Giesbers

46

"Nature is a vital part of European identity. It is a great part of our heritage and a part of ourselves.

Ladislav Miko, Minister of the Environment, Czech Republic

"Wild, undestroyed nature has a value greater to us than many realize. Until it is gone. Then we can really feel the loss of it."

Orsolya Haarberg, Wild Wonders photographer, Hungary

Many of Europe's natural crown jewels are still unseen beyond their own countries. That is hardly surprising. For ages Europe has been divided. East and West were split politically for decades. On top of that, language barriers have separated us. Even our organisation into separate nation-states has meant that much of Europe's nature has gone largely unseen—even by Europeans.

How many western Europeans for example, believe that Mount Blanc is the continent's highest mountain? Well, as you might know, it isn't—not by a long way (sorry about that, France and Italy). At 5,642 m (18,510 ft.), Mount Elbrus, in the wild Caucasus mountain range between Russia and Georgia, is some 800 m (2,700 ft.) higher.

Europe's nature is really much more spectacular and diverse than most of us realise. Landscapes vary from the deserts and salt marshes of Spain in the south, through the dense broadleaf forests of central Europe, to the high Arctic tundra of Spitsbergen and Franz Josef Land in the north; from the lush green Atlantic landscapes of Ireland and the British Isles in the west to the rolling steppes of Russia and Ukraine in the east; and from the coastal wetlands of the Netherlands to the snow-clad peaks of the high Alps, the Carpathians, and the high Caucasus range. With tropical species like bee-eaters, genets, sand grouse, scorpions, and gecko lizards at the one end, and extreme Arctic creatures like polar bears, walrus, and musk oxen at the other.

Cork oak forest
Quercus suber
ITALY/GALLURA, SARDINIA

An enjoyable way to support nature in the Mediterranean is to drink wine from bottles sealed with natural cork stoppers. The cork oak grows for hundreds of years and is the dominant forest-tree species, especially in Sardinia, southern and central Portugal, and Spain. It is vital to a great number of other species there. The harvesting of its bark is done every 10 years or so and doesn't harm the tree. If we switch to plastic stoppers or bag-in-box wines, these forests will disappear.
Staffan Widstrand

Long-legged buzzard bringing a green lizard to its chicks
Buteo rufinus
BULGARIA/NIKOPOL, PLEVEN

A typical steppe species, the long-legged buzzard breeds from the Balkans eastwards into Russia, Turkey, and central Asia. Those that breed in Europe migrate south to Africa and the Middle East for the winter, either passing over the Bosporus or crossing the Caucasus range. It feeds on rodents, lizards, birds, and large insects like grasshoppers.
Dietmar Nill

Lush vegetation around a legendary spring
ALBANIA/SYRI I KALTER

Its name means "the Blue Eye spring," and it feeds Albania's Bistrica river—a green oasis amid Albania's lowlands, which are intensively farmed and logged, and are grazed by goats and sheep.
Anders Geidemark

Wolverine, the hyena of the taiga forest
Gulo gulo
FINLAND/KUIKKA, KUHMO

The wolverine is an animal in need of a PR makeover. Despite its fierce reputation, it is really a shy creature and a quite inefficient hunter. Although widespread, these super-sized martens are uncommon across their range and depend mostly on wolves, lynxes, or bears to provide them with kills to scavenge—a true hyena of the northern forests and tundras. At the same time, wolverines are wary of larger carnivores; as wolves in particular will kill them, given the chance.
Staffan Widstrand

Some of Europe's charismatic species are unseen **because they are so rare.** A few are even on the very brink of extinction. Like the Iberian lynx, one of the world's two most endangered cat species. Or the Mediterranean monk seal, the world's most endangered seal species. Or the beautifully coloured European hamster, which is not as rare, but is rapidly decreasing in numbers. These are all species in trouble. Others are unseen because they live in inaccessible places, like caves or the deep ocean, on mountain tops, in boggy swamps or far-away forests. Others are active mainly at night, like bats, owls, badgers, and martens.

But perhaps surprisingly, many of our amazing wild creatures are within a couple of hours' reach of Europe's main cities; and many of them even live right beside us, in our gardens, parks, and suburbs. If they are unseen, it may be because we haven't bothered to look for them.

On this big chunk of continent, there are many more unseen wild places than most would imagine. Remote areas that are so high, so deep, so dry, so wet, so barren, or so dense that very few of us have ever visited them. It may surprise some of us to hear that real wilderness still exists in Europe. But it's true. Some of these places are so mind-blowingly spectacular, so other-worldly, so somehow sacred that we can do nothing more than stand back and look in awe: The Tichá wilderness in Slovakia; Laponia and Vindelfjällen in Swedish Lapland; Finnmarksvidda, Hardangervidda, and Spitsbergen in Norway; the Rila and Rhodope mountains in Bulgaria; Retezat in Romania; Gran Paradiso in Italy; Vanoise in France; Ordesa in Spain; the Pindos mountains in Greece; Nenetsky, Bryansky Les, and Teberdinsky reserves, and the Volga delta in Russia; almost all of Greenland; and a remarkably long list of others.

So, there is a lot to celebrate in a Europe full of wild wonders. Our natural heritage is still very much out there. Wild wonders big and small. Far away and on our doorstep. Rare and common. Easily seen and elusive. Beautiful, ugly, cool, sexy, warm, cold, high, low, fast, and slow.

We don't always realise it, but Europe's wild wonders are out there serving us all; and they will continue to serve generations yet to come, if we just allow them to.

PAGES 62–63
Basking shark
Cetorhinus maximus
UNITED KINGDOM/ISLE OF MULL, SCOTLAND

The second largest fish species in European waters, the basking shark grows to 12 m (39 ft.) in length and a weight of 19 tonnes (21 tons). It has no teeth but, like the big whales, filter-feeds on small plankton near the surface. Basking sharks are increasingly seen in the waters off southern Britain and the Scottish coast. On the Isle of Mull, you can join organised boat trips to see them or to dive with them.
Nuno Sá

LEFT
Grey heron with a carp catch
Ardea cinerea and Cyprinus carpio
BELARUS/PRIPYATSKI NATIONAL PARK

A true fish specialist, the heron is bouncing back all over Europe now that we've almost stopped killing them. They can live anywhere, as long as there is enough food. Here in the vast Pripyat marshlands, this is not a problem and herons thrive.
Bence Máté

UPPER RIGHT
Bosnian pine
Pinus leucodermis
ITALY/POLLINO NATIONAL PARK, BASILICATA/CALABRIA

The mountain area of Pollino is virtually unknown outside of Italy. Its speciality is the age-old Bosnian pine, an ice-age relict. On their lower slopes, the mountains are covered with vast beech and chestnut forests. The Pollino National Park was created in 1993.
Sandra Bartocha

LOWER RIGHT
Limestone canyon
SLOVENIA/TRIGLAV NATIONAL PARK

The 800 sq. km (309 sq. mi.) Triglav National Park is a strikingly beautiful part of the Julian Alps, a section of the high Alps that stretches down from Austria and Italy. These limestone mountains are riddled with caves, canyons, and waterfalls shaped by rapid rivers that carve through the rock.
Daniel Zupanc

PAGES 66–67
Forest dormouse eating mulberries
Dryomys nitedula
BULGARIA/RHODOPE MOUNTAINS

The forest dormouse is a rodent that is a bit smaller than a squirrel and mostly active at night. It is often in deep sleep during daytime and hibernates in winter. Its taste for grapes makes it unpopular with many wine growers. Bulgaria has the highest number of endangered mammal species of any country in Europe.
Dietmar Nill

PAGES 68-69
Litlanesfoss waterfall
ICELAND/HENGIFOSSÁ RIVER

Iceland has some of the wildest lands on the continent, with huge, virtually uninhabited areas and vast national parks and reserves. Not far from Egilsstadir, there are two jaw-dropping waterfalls. The higher and more famous is Hengifoss; but perhaps even more impressive is Litlanesfoss, just a bit farther downstream. The remarkably regular six-sided columns of volcanic rock, called pillar basalt, were created when lava cooled down over a long period of time.

Orsolya Haarberg

RIGHT
Moss campion
Silene acaulis
LIECHTENSTEIN/MALBUN

Most high-alpine plants face two problems: cold temperatures and lack of water. In response, they have adapted to grow very low and in dense cushions, like this campion. These highly specialised plants are likely to be among the first species to be affected by climate change.

Edwin Giesbers

PAGES 72-73
Roe deer among Siberian iris
Capreolus capreolus and *Iris sibirica*
SLOVAKIA/POLONINY NATIONAL PARK

The roe deer was hunted to near-extinction in many European countries during the 19th century. But thanks to better-enforced hunting regulations, the roe deer is now a common animal again in most of central, eastern, and northern Europe—even taking to our gardens and parks. Its white-spotted fawn is the real Bambi from the original story. The 298 sq. km (115 sq. mi.) Poloniny National Park has been a Unesco World Heritage Area since 2007. It is part of the East Carpathian Biosphere Reserve, which stretches across Slovakia, Ukraine, and Poland.

Konrad Wothe

PAGE 74
The wild coast
SPAIN/ANAGA, TENERIFE, CANARY ISLANDS

The Anaga peninsula is a remote and wild corner of Tenerife, and its misty laurel forests are accessible through ancient hiking trails that have been renovated in recent years. Many species in the Canary Islands, especially among the plants, are endemics, meaning they live only there and nowhere else on Earth.

Iñaki Relanzón

PAGE 75
Emperor moth
Saturnia pavonia
FRANCE/DIGNE LES BAINS,
ALPES DE HAUTE-PROVENCE

These caterpillars have transformed from eggs; they will transform again, to pupae, and finally into moths. Many caterpillars are prickly, stinging, or poisonous—all creatures have evolved their own defence mechanisms.

Niall Benvie

PAGES 76-77
Black grouse cock landing
Tetrao tetrix
SWEDEN/SURAHAMMAR, VÄSTMANLAND

Black grouse cocks gather in spring at 'lek' or display sites, many of which remain the same for centuries. Typically on moors, bogs, or ice-covered lakes, the males display by strutting about, white tails fanned, uttering an other-worldly combination of bubbling and wheezing. The most impressive will get the chance to mate. Their springtime ritual is one of the most evocative sounds of the northern forests.

Erlend Haarberg

PAGES 78-79
Northern lights in the taiga
FINLAND/RIISITUNTURI NATIONAL PARK,
LAPLAND

This could be any mountain spruce forest north of the Arctic Circle. The northern lights, or aurora borealis, are a phenomenon originating in the uppermost part of the atmosphere and caused by solar-wind particles that are speeded up along Earth's magnetic field lines. It happens all year round but is only visible when nights are dark enough and most frequently around the equinoxes, in September/October and March/April.

Sven Začek

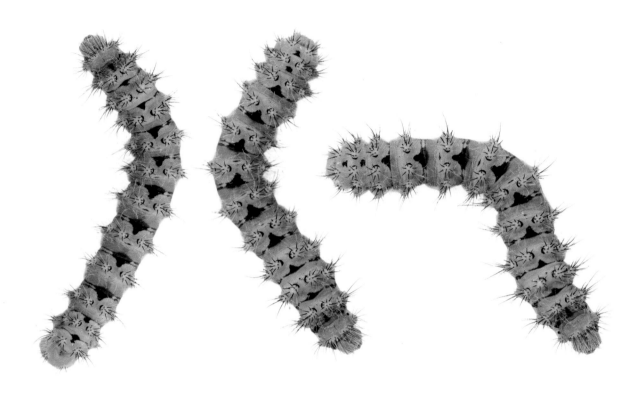

" Just the beauty alone of wildlife is reason enough for us to save it and support it.

Vincent Munier, Wild Wonders of Europe photographer, France

Unexpected

How about some more really good, unexpected news?

A lot of wildlife in Europe is coming back. Yes, you read that right! Not just in one country or two, and not just a few species. After centuries of persecution, carelessness, and wasteful habits, we are now witnessing an almost continent-wide wildlife comeback.

From the majestic eagles, whales, and wolves, to the playful otters, hooting owls, and trumpeting cranes. Side by side with many, many others.

And not only in the distant countryside, but also closer to us, in cities and suburbs, outside factories and on old airfields. Why? Because we have shifted our attitudes. Because Europeans have made a number of decisions in favour of wildlife. Because we don't persecute, harass, and poison wildlife quite the way we used to. This is making all the difference. It is possible to let our wildlife come back. We can do it!

The attack comes completely unexpected from a clear blue sky. The pigeon probably didn't even know what hit it. Struck in a cloud of feathers in midflight by the swiftest predator in the skies, a peregrine falcon. Slate grey above, white underneath, finely barred in black, with an elegantly designed black and white face, intelligent-looking eyes, and a bright yellow base, the bill matchin the deadly feet and talons—the succesful raptor brakes in a steep upward climb, and then turns to catch the lifeless prey as it tumbles to Earth. A classic and timeless wildlife action scene, but this time witnessed by hundreds of churchgoers and tourists, right outside the famous Sagrada Familia cathedral in Barcelona, Spain. An iconic species of our natural heritage, breeding at an iconic site of our cultural heritage. Peregrines have nested here since 2005, high up on the cathedral's famous towers, feeding mainly on the hoards of pigeons that flock here to scavenge offerings from the cathedral's visitors down at street level.

Just two decades ago this charismatic bird of prey was seriously threatened, teetering on the very brink of extinction because of persecution, falconry, and chemical pollutants in many corners of Europe. Now it is making a vigorous recovery. Countries like Sweden have had an almost hundredfold increase in the number of successful breeding pairs during the past 30 years. Peregrines are now colonising one major metropolis after another, seemingly selecting cultural icons as their nesting sites. Today peregrines breed not only on Gaudí's famous cathedral in Barcelona, but also on Notre Dame in Paris, at the Alexanderplatz in Berlin, on the Palace of Culture in Warsaw, and near the London Eye. They can even be seen hunting daily over the pope's residence in the Vatican.

RIGHT
Peregrine falcon
Falco peregrinus
SPAIN/BARCELONA, CATALONIA
The peregrine is considered the fastest bird in the world and has been clocked at more than 300 km/hr (186 mph) when swooping on prey. Today, an estimated 12,000 pairs breed in Europe outside of Russia—a fantastic comeback from the verge of extinction, and the result of decades of focused conservation work. Peregrine eggs in many areas of Europe, however, still have around 100 times higher levels of pesticides than are allowed in the chicken eggs we eat. This shows just how long banned pesticides can remain in the food chain.
Laurent Geslin

Migratory birds belong to all of us, not just to those living where these birds happen to congregate at shooting distance while passing by.

Florian Möllers, Wild Wonders of Europe photographer and director, Germany

This amazing comeback didn't just happen by itself. It is the result of a lot of human effort—research studies, surveillance, protection, captive breeding programmes, reintroduction programmes, tens of thousands of days worked by volunteers, and a lot of public and private funding.

The peregrine falcon is a strong symbol—in older times a symbol of power, agility, hunting ability, sharp vision, and speed. Then in the 1960s it became instead a symbol of the disastrous effects of pollution, a victim of industrial chemicals like DDT and PCBs. But now it is emerging as a symbol of hope, of what is possible when we humans decide to act to sort out the problems we've created. The peregrine is not alone out there in staging a spectacular recovery. It is an indicator of something bigger and even more exciting, a fantastic success story so far almost untold. Europe's wildlife is on the return, right here, right now, and right in front of our very eyes.

RIGHT
Tara River Canyon
MONTENEGRO/DJURDJEVICA, DURMITOR NATIONAL PARK
The 80 km (50 mi.) long and 1,300 m (4,276 ft.) deep Tara River Canyon is another natural wonder of Europe that is almost unknown outside the Balkans. The 320 sq. km (124 sq. mi.) Durmitor National Park has been a Unesco World Heritage Area since 1980. Its highest peak is the Bobotov Kuk at 2,522 m (8,275 ft.).
Milán Radisics

A pan-European wildlife comeback, from north to south and from west to east. The list of bird and other animal species that have increased greatly in numbers over the past 30 years is really impressive. And it is something most of us are completely unaware of. It is unexpected, to say the least.

The otter is coming back again to areas where it was once common. And the beaver is coming back. So are the lammergeier and other vultures. And the sea eagle, big time. And the crane and the whooper swan. And the wolf and the bear and the lynx. And even the wolverine. And the great bustard and the white stork. And the Iberian imperial eagle. And the grey seal. And the orca and the beluga whale and the minke whale.

And even the blue whale, the biggest mammal that has ever existed on earth. And the walrus and the wisent. And the eagle owl, the great grey owl, and the Ural owl. And the arctic fox. And the wild boar. And the spoonbill, the grey heron, the cormorant, the pelicans, and all the egrets. And many river fish species, like the salmon and the brook trout. And almost all species of geese and ducks. And the red deer, the roe deer, the ibex, and the chamois. And lots more besides.

This is something really worth celebrating.

PAGES 86–87
Lammergeiers fighting
Gypaetus barbatus
SPAIN/SIERRA DE BOUMORT, PYRENEES, CATALONIA

Known in Spanish as *quebrantahuesos, the* "bonebreaker", the lammergeier drops bones from a great height on specifically chosen stone slabs in order to crack the bones open and allow it to eat the rich marrow inside. Tortoises are sometimes treated the same way. The lammergeier's gastric juices are pure acid on the pH scale, enabling it to digest large bones. It has the largest wingspan of all raptors in Europe, up to 285 cm (112 in.). Since the 1980s, a series of succesful reintroduction programmes in the Alps has resulted in around 10 new breeding pairs so far. Elsewhere in Europe, lammergeiers can be found in the Caucasus mountains and in the Pyrenees, from which they are slowly spreading throughout Spain.
Magnus Elander

LEFT UPPER AND LOWER
Edelweiss and spring gentian
Leontopodium alpinum
and *Gentiana verna*
LIECHTENSTEIN/MALBUN

Two of the most iconic of all high-alpine flowers, the edelweiss and the gentian are found on several mountains in southern Europe.
Edwin Giesbers

RIGHT
Alpine ibex
Capra ibex
ITALY/GRAN PARADISO NATIONAL PARK, VALLE D'AOSTA

In the mid-1800s there were only 60 Alpine ibex left in this area of Italy. Gran Paradiso was set aside as a royal hunting reserve in 1856 by King Victor Emmanuel II and then made into Italy's first national park in 1920. Now the ibex are thriving, and over the years many ibex from here have been relocated to other areas in the Alps: to Austria, Switzerland, and Germany. Today the Alpine ibex population is estimated at more than 40,000 and still growing.
Erlend Haarberg

There is of course no lack of serious environmental problems all around us. The wildlife comeback is not happening everywhere, and it doesn't cover every species. Some countries are still plagued by widespread illegal hunting and fishing. There are severe problems caused by modern forestry and farming, and there is a disaster happening in our seas due to massive overfishing. We hear about these and other man-made environmental problems day and night, and they definitely need to be dealt with. But while doing that, the inspiring and positive wildlife comeback can really teach us something—something that gives us all hope. Our societies have deliberately chosen, at least recently and at least partly, to start changing their attitude towards wild things. And that really does make a difference.

This is a success born from active and informed nature-conservation work. Behind every species-comeback story, there are devoted conservationists. Other steps, large and small, have been taken by national governments, the European Union, the United Nations, businesses, nongovernmental organisations, landowners, and individuals, which have all mounted up to an impressive overall nature-conservation effort. Networks of new and better protected areas, the banning and phasing-out of a number of really vicious chemicals, the introduction and enforcement of more civilized hunting laws that are more ethically and ecologically sound, reintroduction and stricter protection measures for a number of species. These are some of the reasons why these places and species are still with us and why wildlife is returning. So long as a

PAGES 90–91
Monte Paterno and Tre Cime di Lavaredo
ITALY/DOLOMITES, SÜDTIROL, BOLZANO

This area of the Dolomites is famous with hikers worldwide, and the region was declared a Unesco World Heritage Area in 2009. Cima Grande, the beautiful peak in the middle, is the highest, at 3,000 m (9,840 ft.).
Frank Krahmer

LEFT
Black vulture
Aegypius monachus
SPAIN/MONFRAGÜE NATIONAL PARK, EXTREMADURA

Following centuries of persecution, the huge Eurasian black vulture is now coming back in Spain and France, due largely to the feeding of vultures at specific sites. Overzealous EU veterinary regulations mean there are fewer dead animals left in the countryside today, and this has created a disaster for all vulture species. They are also suffering from shepherds putting out poisoned carcasses to kill wolves. In Greece and the Balkans, black vultures are therefore still declining. The world's greatest concentration of black vultures nest in and around the 195 sq. km (75 sq. mi.) Monfragüe National Park, which was designated in 2007.
Markus Varesvuo

RIGHT
Wild boar sow
Sus scrofa
UNITED KINGDOM/ALLADALE WILDERNESS RESERVE, SCOTLAND

This ancestor of our domestic pig is one of the most widespread and familiar of all European animals. Despite their reputation and size—in Russia and Romania, 300 kg (660 lb.) boars have been found—these are generally shy, peaceful animals and are quietly finding their way back to places where they've not been seen for years. Extinct in Britain since the 13th century and in Scandinavia since the 14th, boars are now reestablishing themselves in both regions, mainly through private reintroductions and mass escapes from boar farms. Not everyone is pleased about their resurgence. Gardeners and farmers in particular know that a family of boars can ruin years of effort overnight. But on the plus side, boars are great rooters and snufflers and are perfect agents for forest regeneration.
Peter Cairns

species is not completely lost and extinct, there is hope. Of course the rarer a species becomes, the more expensive it is to bring it back. Successful reintroductions so far include red kites in Wales and England; sea eagles in Scotland; lammergeiers, griffon vultures, and ibex in the Alps; peregrine falcons in many countries; large blue butterflies in Britain; lynxes in Poland and Switzerland; red deer in Corsica; great bustards in Britain and Germany; beavers and wild boars in Sweden; sturgeons in Poland; and arctic foxes in Norway. Plans are also underway for restoring species as different as monk seals, sea turtles, Iberian lynxes, wolves, bears, and eels.

RIGHT
Kingfisher success
Alcedo atthis
HUNGARY/LAKE BALATON, BALATONFŰZFŐ, VESZPRÉM

The widespread Eurasian kingfisher breeds from northwest Africa to Japan, and in most of Europe north to Scandinavia. A fish specialist that also eats insects, it digs its nesting tunnel into steep waterside sand banks, where it lays five to ten eggs. It often has two broods in a season, sometimes even three. Cold winters however, can take a great toll on local populations. The kingfisher is benefiting from many of Europe's freshwater systems being much less polluted than they were just 20 years ago.
László Novák

PAGES 96-97
Berger's clouded yellow butterfly
Colias alfacariensis
LIECHTENSTEIN/MALBUN

This butterfly species is widespread across central Europe and the Mediterranean. It is very local, rarely moving more than a kilometre (less than a mile) from its birthplace. Many butterflies are among the losers in the European biodiversity game. They are all strongly connected to wildflowers. Since flowering meadows are disappearing at a rapid rate, many butterfly lovers have started to spread wildflower seeds along roadsides and in public parklands.
Edwin Giesbers

PAGES 98-99
Gladiolus
Gladiolus palustris
LIECHTENSTEIN/MALBUN

One of many wild gladiolus species in southern Europe. Decorative for humans to look at, but a necessity of life for many butterflies and other insects, this is the wild origin of the domestic gladiolus that we now see sold in garden centres and supermarkets.
Edwin Giesbers

Even though Europeans have voted for better laws and policies towards our natural heritage, there is still a lot to be done. We have more scientific knowledge about nature today, and we are perhaps less persuaded by ignorance and prejudice than in times gone by. Yet in some areas old habits die hard, and the barbaric trapping of songbirds, the poisoning of wolves, the illegal mass slaughter of saiga antelopes, and the sport shooting of migrating raptors still persist. But luckily, these habits seem to be dying out. They represent the past, not the future, not the youth, not modern ways. One day we will look upon this behaviour with the same bewilderment as we now look upon witch-burning or slavery.

Our bewilderment will also extend to the massive government subsidies—at both the national and the EU level—that have supported the pointless destruction of wildlife and unique habitats.

It is clear that when we treat our fellow species with just a little more respect and tolerance, we are immediately rewarded with their return. Imagine then what would happen if we chose to take another small step or two, allowing species and ecosystems just a little more leeway and opportunity to do their thing undisturbed?

How amazing and rewarding would that be?

Not only for the wild beings around us, but also for ourselves.

LEFT
Bee-eater tossing a bumblebee
Merops apiaster and *Bombus sp.*
HUNGARY/PUSZTASZER PROTECTED LANDSCAPE, CSONGRÁD

One of Europe's most exotic-looking birds, the bee-eater lives in colonies in sandbanks. As such, this species has benefited from human construction and road building that provide many more artificial sandbanks than in untouched nature. The bee-eater is a Mediterranean species, spreading northwards with climate change. Sometimes they are persecuted by bee-keepers.
Markus Varesvuo

UPPER RIGHT
Turquoise European rollers
Coracias garrulus
HUNGARY/PUSZTASZER PROTECTED LANDSCAPE, CSONGRÁD

No one ever forgets a close encounter with a roller and its elegant mix of turquoise and cobalt blues shining in the sun. It eats mainly large insects, and it nests in natural tree holes. As both are declining, so is the roller; such are the connections in nature. Artificial nest boxes have been shown to help, at least with the lodging problem.
Markus Varesvuo

LOWER RIGHT
A thrush's egg
Turdus sp.
LUXEMBOURG/LA PETITE SUISSE LUXEMBOURGEOISE, MULLERTHAL

Like a jewel on the forest floor, this egg had fallen out of a thrush's nest, perhaps a cuckoo chick threw it out together with all the other eggs from the nest.
Jesper Tønning

PAGES 102-103
Polar bear in front of glacier ice wall
Ursus maritimus
NORWAY/SVALBARD

Norway and Russia have huge protected areas and sizeable polar bear populations. So do northern Greenland and the pack ice in between. Polar bear numbers increased when commercial hunting was banned in 1973 and have now stabilized at about 25,000 worldwide. The ice bear now faces another challenge: that of climate change. This bear was photographed on Svalbard where ship-bound polar bear safaris provide a great opportunity to see this icon of the north.
Ole Jørgen Liodden

"The Mediterranean basin is one of the world's 25 top Biodiversity Hotspots. This fact demands a great responsibility. Are we all taking that seriously enough?

HSH Prince Albert II of Monaco

We can actually live alongside a lot of wildlife species. We can even harvest from their abundance, if we stick to nature's rules. It is almost entirely a question of what attitude we want to take towards nature. We can live beside wildlife even in big cities. In the high-rise suburb of Racadau in the city of Brasov in Romania, there are brown bears living right outside the houses in the nearby forest. They walk the streets at night, with the 25,000 people living there seemingly used to and even proud of their bears. They are learning how to live beside these powerful predators. Of course, it is not without problems; but you meet families out bear-spotting at night on foot, kids and parents holding hands and walking by the bear families at quite close range. In Brasov, wolves are also denning; and they cross right through the centre of the city every morning and evening on their way to and from their feeding grounds—the local garbage dump. They feed on the rats that live there. Or take the city of Berlin, where some 5,000 wild boars have settled; or downtown London or Paris, where foxes den; or a city like Helsinki where eagle owls breed on rooftops in the city centre.

Quite unexpected.

PAGES 110–111
Balkan waterfalls at Lake Milanovac
CROATIA/PLITVIČKA JEZERA NATIONAL PARK

Calcium carbonates from thousands of years of rushing water dissolving the brittle karstic limestone rock, together with algae and moss, create low walls of tuff or of travertine between the lakes and in the waterfalls. Plitvička has been a Unesco World Heritage Area since 1979 because of its outstanding, almost surreal beauty.

Maurizio Biancarelli

PAGE 112
Great horsetail
Equisetum telmateia
SLOVAKIA/POLONINY NATIONAL PARK

The horsetail is a living fossil, the only remaining genus from a group of plants that were common 100 million years ago. Some species back then were large trees, 30 m (98 ft.) tall, which later turned into carbon and are an important part of all fossil coal deposits worldwide. Several horsetail species of today have a coarse skin and have been used to smooth and polish fine woodwork, and to clean pots and pans.

Konrad Wothe

PAGE 113
Lesser horseshoe bat
Rhinolophus hipposideros
ITALY/GROTTA DI MONTE MAJORE, SARDINIA

The lesser horseshoe bat is one of the world's smallest bats. It weighs 5 to 9 g (0.2 to 0.3 oz.) and its pups only 1.8 g (0.06 oz.) at birth. Bats often gather in caves for the winter, like this limestone stalagtite cave on the island of Sardinia. Sardinia is yet another place where habitat and species restoration is succeeding in bringing back species from the brink, like the mouflon sheep, the golden eagle, the red deer, and the lammergeier.

Ingo Arndt

RIGHT
Alpine marmot
Marmota marmota
AUSTRIA/HOHE TAUERN NATIONAL PARK, TYROL/KÄRNTEN/SALZBURG

Marmots are sleepy creatures that hibernate for up to nine months each year. But when they wake, they work very hard, digging huge burrow systems in the ground. Weighing 8 kg (18 lbs.), they are the largest species of the squirrel family. They live in colonies in the Alps, Carpathians, Tatras, Pyrenees, and Apennines. Hohe Tauern National Park, at 1,834 sq km (708 sq. mi.) is the largest nature reserve in the Alps, nominated to become a Unesco World - Heritage Area.

Grzegorz Leśniewski

"A hole in our souls. That is what we are left with when a species is lost.

Kemal Nuraydin, Executive Editor, *National Geographic Turkey*

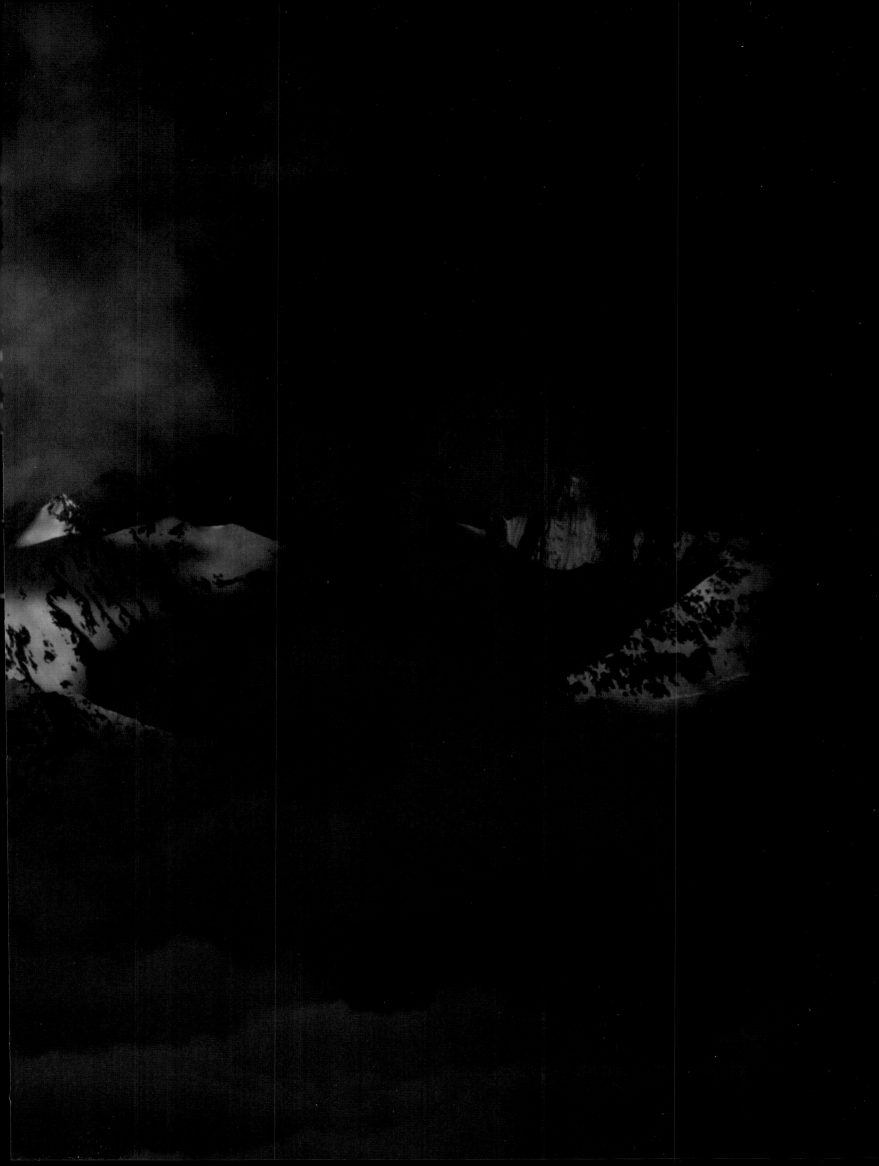

Unforgettable

From the Altamira cave painters, Greek mosaic layers, and Viking rock carvers, to today's architects, musicians, and designers, nature has always been a huge source of inspiration. The same is true for most of us, whether we realise it or not. We have a need to connect with nature around us and to be soothed and healed by it. That is no New-Age theory, just practical, ancient knowledge. Nature simply helps us to set our priorities straight. We need to have fun in the great outdoors and to feel the wow factor of meeting wild creatures and being in wild places. Many of Europe's most unforgettable wildlife encounters are also now becoming easier for us to have than ever before. We can even book many of them! Tourism is bringing new economic value to nature at the same time that it makes nature more accessible to many of us.

Suddenly we are fully awake. It sounds like someone is walking by in the bog with heavy splashing strides outside the little plywood hut we're in. Strange, this is in a no-man's-land right on the border of Russia and Finland; and there shouldn't be anyone walking around out there. But, hey, that sounds rather like four legs, not two. When we peek out into the pre-dawn light, the shadow of a huge bear appears, calmly passing by the hut, just 5 m (16 ft.) away. Farther off in the bog, something else is happening; but it's difficult to see exactly what. Deep bear growls and grunts, combined with wolf yelps, give us a hint. Half an hour later, there is light enough to see the shapes of no fewer than nine bears and eight wolves dodging around each other in the misty dawn.

The wolves are trying to fool the bears and then steal some of their food, a big moose carcass. The bears, on the other hand, are determined to protect their meal. This ancient battle between wolf and bear over their prey has been waged all over Europe since the beginning of time. Today it is extremely rare to be able to witness this anywhere in the world; but here in Kuhmo, on Finland's border with Russia, it is possible. Not only because there are bears and wolves in these forests, but also because there are now carnivore-watching facilities, this one run by the Finnish Wild Wonders photographer Lassi Rautiainen. Later on in the morning, after all the bears and wolves have left, and we are just about to leave Lassi's hut on foot, we suddenly see two wolverines lolloping through the forest towards the scattered remains of the carcass. So we stay on for another hour, not wishing to disturb them during their breakfast.

"Designating an area as a national park usually brings a great increase in nature tourism, providing jobs and income for people who live there.

Zoltán Kun, Executive Director, PAN Parks Foundation, Hungary

Brown bear
Ursus arctos
FINLAND/MARTINSELKONEN, SUOMUSSALMI
Ten years ago, most Europeans who wanted to see bears went to Alaska. Now they go to Finland, Sweden, or Romania, which all have large and growing bear populations and tour operators who will take you to view the bears from permanent huts.
Peter Cairns

PAGES 132–133
Wild tundra reindeer
Rangifer tarandus
NORWAY/FOROLLHOGNA NATIONAL PARK, HEDMARK/SØR-TRØNDELAG
Wild tundra reindeer are today confined to Hardangervidda and Forollhogna in Norway and in some places in arctic Russia. There are also small groups of wild forest reindeer in the borderlands between Finland and Russia. All other European reindeer are owned by someone, even if they run freely in the mountains.
Vincent Munier

PAGE 134
Griffon Vulture
Gyps fulvus
SPAIN/PARQUE NATURAL DE LAS HOCES DEL RÍO RIAZA, CASTILLA Y LEÓN
In Spain, the comeback of the griffon vulture is a great success—from 7,000 pairs in 1980 to around 18,000 in 2009. Griffons now breed in 16 European countries and have recently been seen in Belgium, Germany, and the Netherlands. About 1,000 vultures are killed each year in Spain through collisions with wind turbines.
Staffan Widstrand

PAGE 135
European Yellow Scorpion
Buthus occitanus
SPAIN/ALICANTE
Common animals in drier parts of the Mediterranean region. Nighttime predators, mainly hunting insects, caterpillars, and spiders, they rest up during the heat of the day under stones. Recent research shows that their strong neurotoxic poison might be used for drugs against epilepsy. One never knows when or how biodiversity will help us next.
Niall Benvie

"The cost of inaction has a much higher price tag than action.

Jacqueline McGlade, Executive Director, European Environment Agency, Denmark

This unique place is one of the very few in the whole world where one is quite likely to see bears, wolves, and wolverines. Sometimes even on the same day and at quite close quarters. But it has taken almost 25 years to build a relationship of trust with the carnivores here. Normally shy and extremely wary, these animals have decided to be a little bit more relaxed towards people. Lassi's place is the only one offering a good possibility of seeing all three carnivores, but there are several other bear-watching sites, both in Finland and Sweden. Some are also being developed in Romania. During 2008, some 4,000 people went bear-watching in one of the 12 facilities in Finland, each person staying an average of four nights. That makes about 16,000 nights, at around €200 per person. Not many other legal businesses in these very remote forests bring in amounts like that on a sustainable basis. The bear has suddenly become a money earner and job creator in these rural areas. And it has become the tourism icon for the whole region, possibly even for the entire country.

Taiga autumn
FINLAND/OULANKA NATIONAL PARK, LAPLAND

The largest intact areas of ancient forest can be found in Russia, Romania, Finland, Norway, and Sweden. These are havens for many species of fungi, lichens, insects, and plants, very rare or completely absent in younger forests. Within the EU, old-growth forests today represent less than 1 per cent of the productive forest area.

Staffan Widstrand

UPPER RIGHT
Wild wolf family visit
Canis lupus
RUSSIA/VORONOVO, SMOLENSK

A unique picture of a female wolf and her cubs outside their den. The legendarily shy wolf is slowly finding its way back to central Europe; even Germany had some 45 wolves in 2009. Russia had possibly 40,000; Romania 3,000; Spain 2,500; Italy 600 to 800; and many other countries 100 to 300 each. Many love it, others hate it, few know anything about it; but everyone has an opinion. The genetic variation within the wolf has given rise to every single dog breed we have today, from Chihuahuas to Great Danes. That is the power of biodiversity.

Sergey Gorshkov

LOWER RIGHT
Brown bear watching
Ursus arctos
FINLAND/KUIKKA, KUHMO

With bears slowly coming back in numbers, some people worry they might be dangerous. Extremely few people are ever attacked by bears, even where bears are common. Reported attacks are almost always a result of an encounter with hunting dogs or of the bear being shot and wounded. Women and children are virtually never attacked. Finland had around 16,000 bear-watching tourist nights in 2009, and not one attempted attack was reported. That said, one should never approach a bear, and especially not small cubs, as mother bears are known to be very protective.

Staffan Widstrand

PAGES 138–139
Classic northern bog
LATVIA/KEMERI NATIONAL PARK

The Kemeru Tirelis bogs, lakes, and mires make up most of this 381 sq. km (147 sq. mi.) national park, created in 1997—one of four in Latvia. The white dots in the foreground are cotton grass, which blooms in July and August.

Diego López

A huge sea eagle comes down from the sky with its fierce yellow eyes fixed on a mackerel near the sea's surface. It swoops in over the water and seizes the fish from the fjord with a swipe of its finger-sized talons. There is a big splash, and then lunch is already on its way to the eagle's chick. Another unforgettable sight usually witnessed by very few and from a long way off. But here in Flatanger, Norway, scenes like this can be seen 20 times a day, right beside your boat! An unforgettable experience, made possible by Ole Martin Dahle, the "Eagle Man", who has been feeding sea eagles here for 20 years, gaining their trust in the process. In winter he feeds golden eagles on land instead, serving them roadkill from the region, and providing reasonably comfortable huts for anyone who wants to see them at close quarters. Meeting the soul-piercing gaze of a wild and free golden eagle from only a few metres' distance is something that stays with you for life.

In the Pyrenees in northern Spain, you can meet the lammergeier, one of the world's most impressive birds of prey. Local authorities and bird-watching tour companies offer huts at vulture feeding stations; and apart from lammergeiers, you can also expect to see hundreds of griffon vultures.

Hungarian Wild Wonders photographer Bence Máté has taken the concept even further. With an extensive system of huts on the Hortobágy Puszta and the Pusztaszeri nature reserve in Hungary, he offers close encounters with a number of the most exotic-looking birds in Europe—hoopoes, bee-eaters, rollers, red-footed falcons, and spoonbills, as well as a variety of egrets, storks, herons, and small birds. Another Wild Wonders of Europe photographer, Peter Cairns, offers something similar in Scotland's Cairngorms National Park, getting up close and personal with red squirrels, red deer, ospreys, and other Scottish icons. And on the hot steppes of La Serena in Extremadura, Spain, German nature-tour operator Joachim Griesinger offers close contact with great bustards, white storks, hoopoes, lesser kestrels, and little owls, in addition to close encounters with hundreds of vultures, in the Parque Natural de las Hoces del Río Riaza in Castilla y León, Spain.

You can also be taken on horseback into the high Caucasus Mountains in Russia and by boat deep into the Danube delta in Romania. You can join steppe tours in Ukraine; take underwater photography dives in the Russian Barents Sea; swim with beluga whales in the Russian White Sea; dive with huge groupers in the Lavezzi reserve in Sardinia, Italy; or swim with wolf fish, angler fish, and halibut at Saltstraumen, Norway. Visit a gannetry on the Bass Rock in Scotland, visit the Outer Hebrides and St Kilda seabird cliffs, or be taken to meet 100,000 geese at Durankulak Lake in Bulgaria. Why not hike in the laurel forests of La Palma in the Canary Islands, Spain? Go whale diving in the Azores, Portugal; or whale watching at Andenes, Norway. You can even swim with killer whales in Møre, Norway. Walk in the Alps with ibex and chamois around you in the Gran Paradiso National Park, Valle d'Aosta, Italy. See walruses and polar bears up close on Svalbard, on specially designed photography trips with Wild Wonders photographer Ole Jørgen Liodden. Meet hundreds of grey seals on the beach at Donna Nook, England; or help feed the pelicans from a local fishing boat in Lake Kerkini, Greece. Let Wild Wonders photographer Jari Peltomäki show you the great grey owls, Siberian jays, or fishing ospreys in Finland. Meet tens of thousands of cranes as they stop over during their migration at Montier-en-Der in France or at Rügen in Germany. Or rent huts by the black grouse or capercaillie leks in Swedish Bergslagen. Fish

Bicaz Gorge is one of several fabulous Romanian national parks in the Carpathian Mountains. These wild refuges—in parks like Retezat, Piatra Craiului, and Apuseni—are home to high numbers of deer, chamois, wolves, and bears. But reckless logging inside some of the national parks is a rapidly growing problem.
Cornelia Dörr

LEFT
Otter having an eel for a snack
Lutra lutra and *Anguilla anguilla*
UNITED KINGDOM/RIVER TWEED, NORTHUMBERLAND, ENGLAND

The otter is still another comeback species in Europe. With bounties paid for them, hunted for their fur, and finally hit by toxic pesticides, the otter disappeared from much of Europe. Now these playful fish-eaters are rapidly returning to many major river systems and along Europe's coasts, thanks to hunting bans and the phasing-out of the worst pesticides.
Laurie Campbell

UPPER RIGHT
Humpback whale fluking
Megaptera novaeangliae
ICELAND/SKÁLFJANDI BAY, HUSÁVIK

The humpback is found in all oceans, migrating 25,000 km (15,530 mi.) or more between its summer hide-outs in the Arctic or the Antarctic and its wintering areas in the tropics. Humpbacks declined by over 90 per cent because of industrial whaling. Since that was banned in 1966, the humpbacks have bounced back. Income from whale-watching tourism in Iceland is now more than twice that of the country's whaling industry at the time it was banned. Still, Iceland took up commercial whaling again in 2009.
Mark Carwardine

LOWER RIGHT
Grey seal battle
Halichoerus grypus
UNITED KINGDOM/DONNA NOOK, LINCOLNSHIRE

A seal watcher's paradise. The Royal Air Force (RAF) bombing range at Donna Nook on the Lincolnshire coast is temporary home to over 3,000 grey seals who come ashore each winter to give birth to their snowy white pups and for the males to fight over the right to sire the next generation. On weekends, when the RAF have time off, thousands of people flock here to get up close and personal with a true wildlife spectacle.
Laurent Geslin

catch-and-release for huge salmon in the Orkla river, Norway. Check out 56 species of orchids in Monte Gargano, Italy; or meet 20 moose in a night at Skinnskatteberg, Sweden; and learn how to attract them by sounding like a moose cow in heat. See hundreds of red kites at Gigrin Farm in Wales; stalk magnificent red deer at Alladale in Scotland; hike among the last population of wild tundra reindeer in Forollhogna; or meet the musk oxen on Dovrefjell, in Norway. These are all bookable nature experiences.

RIGHT
Short-finned pilot whale
Globicephala macrorhynchus
PORTUGAL/AZORES
The pilot whale is a very sociable kind of dolphin, sometimes referred to as the "cheetah of the deep" for its high-speed hunting of deep-sea squid. It lives in big pods of tens or even hundreds, and as a consequence is the only animal known to displace orcas, which seem to panic in the presence of pilot whale pods.
Magnus Lundgren

PAGES 152–153
Buttercup fields
Ranunculus sp.
ITALY/MONTI SIBILLINI NATIONAL PARK, UMBRIA/MARCHE
You may have noticed that buttercups are much more visible in grazed meadows. This is because, although beautiful, they are poisonous to livestock and are therefore usually left to grow in peace. Once dried and turned into hay, they lose their poison and are perfectly edible.
Sandra Bartocha

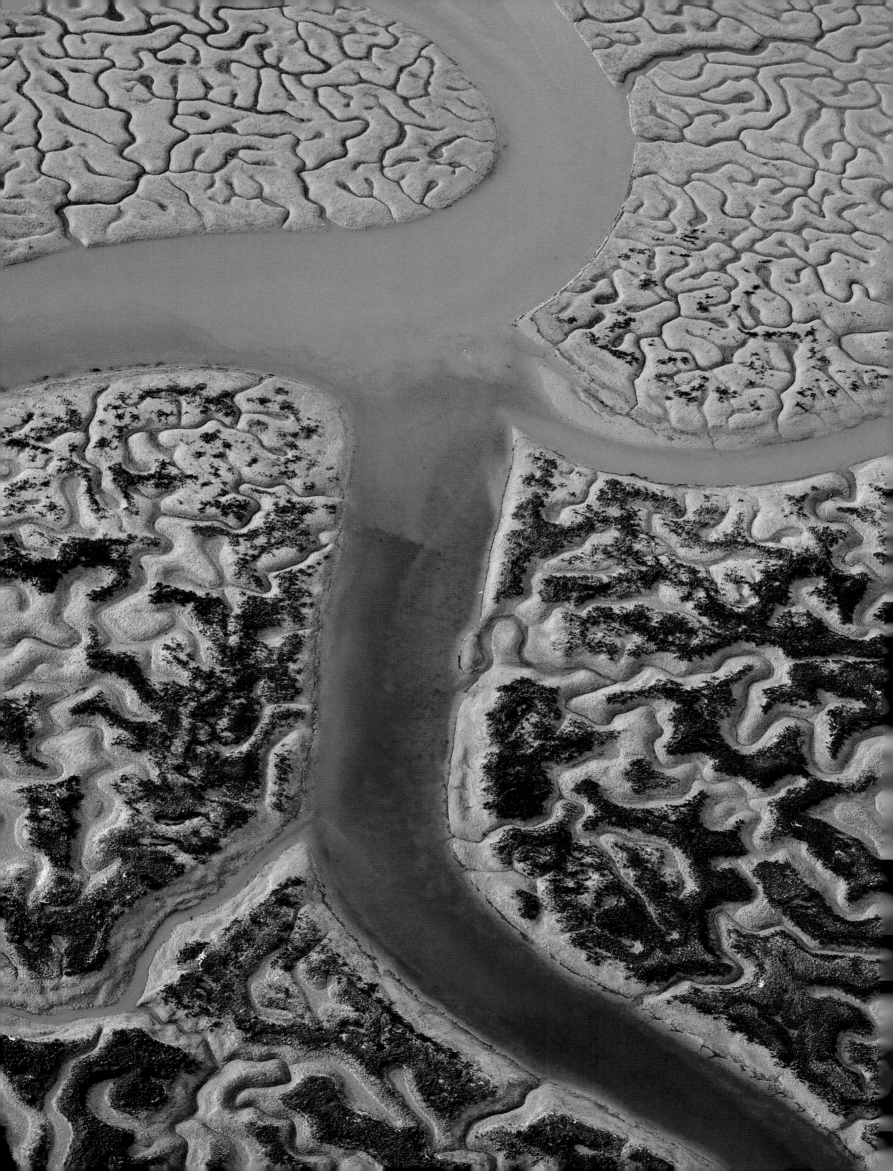

"We should not limit our attention to protected areas only, because these then risk becoming islands of nature in a sea of destruction. Biodiversity is crucial to our well-being.

José Manuel Barroso, President of the European Commission, Portugal

Nature tourism is developing very quickly. In order for it to become an asset rather than yet another problem, it needs to be carried out in an intelligent way, following certain guidelines. This is often referred to as ecotourism, and ecotourism standards are now being developed in several countries. Australia was first with a nationwide quality-labelling system, but Europe soon followed suit. The most ambitious so far are Sweden's (2002) and Romania's (2006). Estonia's was the first (2000) and Norway's the latest (2007). An organisation called the PAN Parks Foundation has also connected a number of wilderness national parks in many European countries and is working to make these areas profitable for the people living there, through tourism.

When we spend more time in nature, we benefit from it ourselves. It is about our own personal well-being and happiness. If we then book certified lodging and tour products, we help bring business and jobs to the area; at the same time we help support nature-conservation efforts there. For the first time in history, unexploited nature is beginning to have an economic value. From old farmland and forests to pure wilderness areas, these wild places used to be regarded as wasteland that produced nothing worthwile. Wildlife was considered useful only if it could be eaten or shot. Instead, the wild places now have a new value.

Protected areas in Germany today have about 290 million visitors a year. The economic and political dimensions of this are something that no politician or corporate director can afford to dismiss in the long run.

PAGE 154
Salt marsh mosaics
SPAIN/PARQUE NATURAL BAHÍA DE CÁDIZ, ·ANDALUCÍA

Primeval mudflats shaped by the twice-daily movements of Atlantic tidal waters. Green algae grows in the zone between high and low tides. All around this oasis there is now industrial development, highways, and harbours. But this jewel is still here and is one of over 25,000 protected areas in the Natura 2000 network. The Spanish coasts are some of the most heavily exploited in all of Europe, and the last remaining wild stretches are in urgent need of protection.
Diego López

PAGE 155
Grasshopper
Miramella alpina
AUSTRIA/FLIESSER SONNENHANG, TYROL

Old farming landscapes have many more flower, insect, reptile, and bird species than can be found in industrial farming areas. A variety of life that has developed over thousands of years. The sun-drenched mountain meadows here in Naturpark Kaunergrat have the highest number of butterfly species in Europe. And some pretty awesome grasshoppers too.
Niall Benvie

LEFT
Pyrenean brook salamander
Calotriton asper
ANDORRA/RANSOL

This is one of Europe's specialised salamander or newt species. It exists only in the Pyrenees and the northernmost Spanish highlands, and nowhere else in the world. These salamanders are very vulnerable, threatened mainly by construction of ski lifts and resorts destroying their breeding waters, and by the introduction of trout to previously trout-free highland rivers and brooks.
Magnus Elander

RIGHT
Marsh helleborine
Epipactis palustris
DENMARK/HØJE MØN

At 60 cm (24 in.) tall, the marsh helleborine is a very visible orchid. It grows in marshes and wetlands on limestone soils, preferring shady spots, during July and September.
Sandra Bartocha

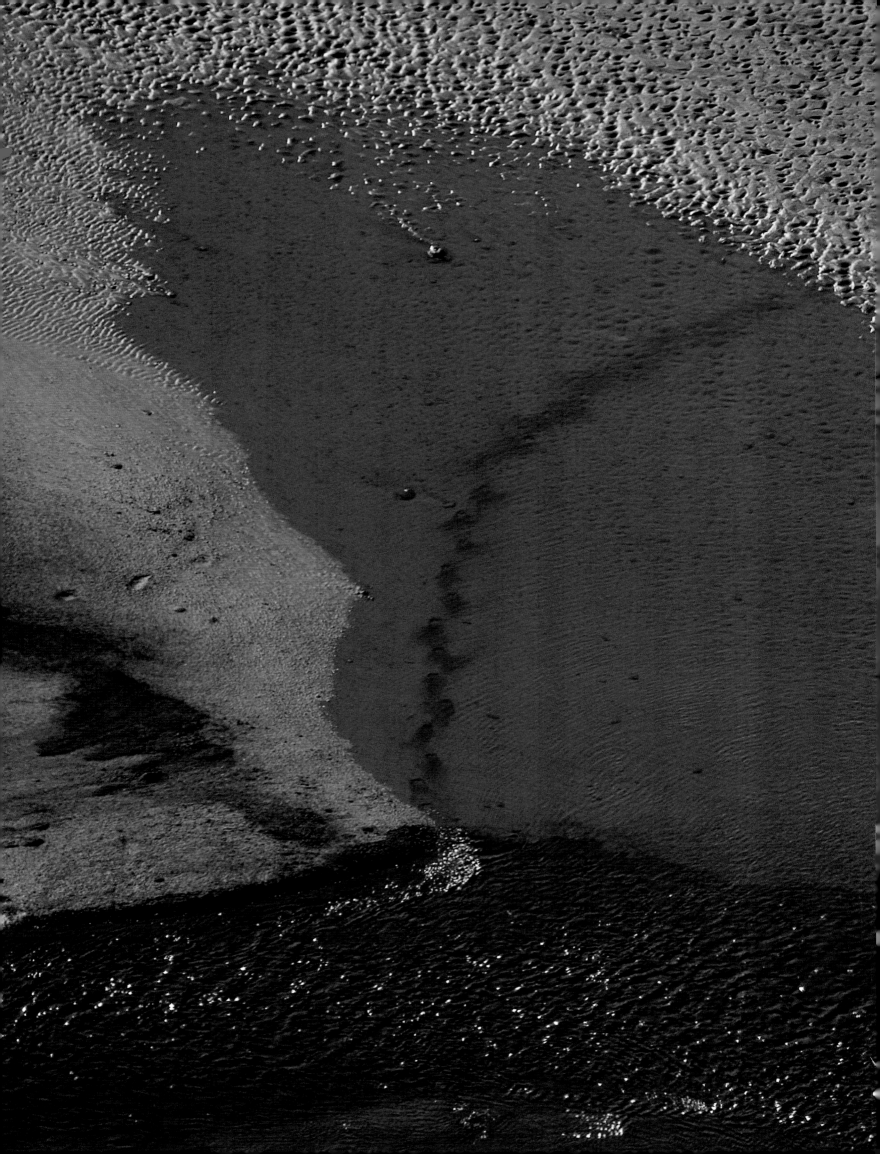

The wild may be priceless; but it also means business, good business. Real, undestroyed wilderness will very soon become the most precious asset of all, with rising value over centuries to come. Wilderness means land that is kept free from roads and powerlines, land that is free from intensive farming, industrial forestry, and construction work. No hydroelectric dams, no wind farms, no highways. Land almost without people or with people who live lightly on it. Land that just is. Land that represents the very core of our shared natural heritage.

We need to be able to enjoy such land, to revere it, to reconnect with it, to be soothed and strengthened by it; and we need to take better care of it and protect it. We must also fight for it. Because no matter what we might think in the short term, without these unforgettable wild lands and wild beings, we are less. Without them, we are diminished as human beings—as individuals, as communities, and as Europeans.

So a call to all parents: bring your kids to Europe's wild areas, enjoy meeting our wildlife, and have fun out there.

You are all in for something unforgettable.

PAGES 158-159
Laponian elk or moose
Alces alces
SWEDEN/SAREK NATIONAL PARK, LAPLAND

Scandinavia has the densest elk population in the world. After they were hunted almost to extinction in the 1800s, the elk hunt became strictly regulated. There is now a sustainable annual harvest of up to 170,000 in Norway, Finland, and Sweden. That is more than in all of North America. Elk-watching safaris are also increasingly popular.

Peter Cairns

PAGES 160-161
Black woodpecker
Dryocopus martius
FINLAND/OULANKA NATIONAL PARK, LAPLAND

One of the biggest woodpeckers in the world, this very vocal bird inhabits most of Europe's northern forests. When you find huge holes in seemingly healthy trees, it is usually the work of the black woodpecker, having used its chisel bill to reach the big ants that live inside the wood.

Sven Začek

LEFT
Great snipe displaying
Gallinago media
ESTONIA/MATSALU NATIONAL PARK

Matsalu Bay is a vast wetland area of over 400 sq. km (150 sq. mi.) where the Kasari River delta meets the Baltic Sea. It is visited by around 2 million migrating ducks, swans, and geese every year, on their way to or from the Arctic tundra. It is also a breeding area for a wealth of wildfowl and wader species, one of which, the great snipe, has organised display sites or leks. The great snipe returns to these leks year after year and can be found within 1 m (40 in.) of where it was the year before.

Lassi Rautiainen

RIGHT
The Rapa River delta
SWEDEN/SAREK NATIONAL PARK, LAPLAND

Sarek National Park is one of the real wilderness icons of our continent, often called the "Alaska of Europe". Together with the neighbouring national parks of Padjelanta, Muddus, and Stora Sjöfallet, and the huge wetland nature reserve of Sjaunja, it constitutes the Laponia Unesco World Heritage Area. At 9,400 sq. km (3,630 sq. mi.), it is one of the largest protected areas in Europe, a continent where only 1 per cent of the land area is considered wilderness.

Peter Cairns

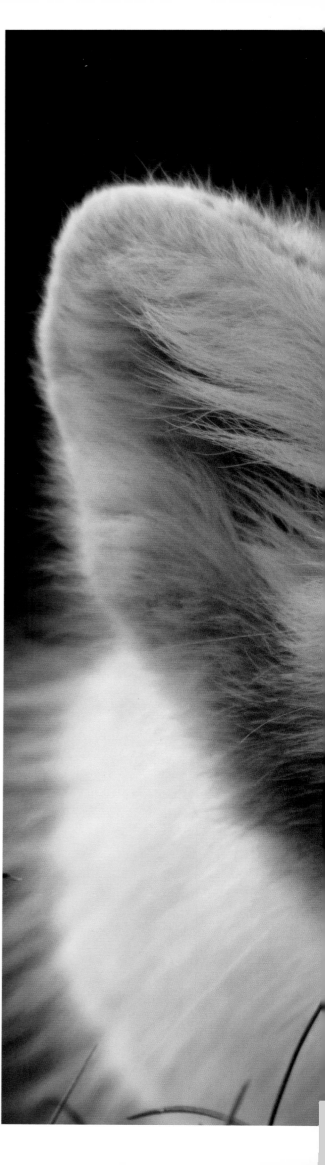

Aegean sunset
GREECE/ALONISSOS NORTHERN
SPORADES MARINE NATIONAL PARK

With its 2,260 sq. km (870 sq. mi.) area,
Alonissos is one of the largest marine
reserves in the Mediterranean. Initially
created in 1992 to protect one of the
last strongholds of the endangered
monk seal, Alonissos is also home to
the unique Eleonora's falcon that
breeds here in several colonies.

Stefano Unterthiner

Azure damselfly
Coenagrion puella
MACEDONIA/LAKE OHRID, GALICICA
NATIONAL PARK

Damselflies are often coloured in bright
turquoise and blue. They are potent
predators of small insects and spiders,
always found near water. Galicica
National Park was created in 1958
between lakes Ohrid and Prespa, and it
is now a part of the transnational park
across the borders of Greece, Albania,
and Macedonia designated in 2000.

David Maitland

Dew drops on grass
SAN MARINO

Grass grows almost everywhere, but
how many of us think about all the
different varieties and species? Many of
our most important food crops, like
wheat, barley, oats, rye, and rice,
originate from the grass family. And
grass is the staple food for most of our
livestock. From back when we were still
hunter-gatherers, grasses have always
been crucial to human survival.

Florian Möllers

Arctic fox
Alopex lagopus
NORWAY/SVALBARD

The arctic fox is an opportunist that
eats almost anything but specialises in
small rodents and birds, when it has
a choice. It is a common animal in
Greenland, Iceland, Svalbard, and
the Russian arctic. In Finland and
Scandinavia it was driven close to
extinction by hunting and trapping for
its valuable fur. Despite over 75 years of
protection, the Scandinavian population
still remains at the brink of extinction,
with only around 200 individuals in
the wild.

Mireille de la Lez

Dice snake
Natrix tessellata
GREECE/PATRAS, PELOPONESSUS

The dice snake is a nonpoisonous
snake. It spends as much time in water
as on land. It swims well and hunts for
small fish, frogs, and tadpoles in lakes
and streams. Many people are scared of
snakes, but very few snakes can
actually do people any harm—and those
that can, rarely do.

Christian Ziegler

Common scorpion fly
Panorpa communis
AUSTRIA/FLIESSER SONNENHANG, TYROL

Scorpion flies look like they have a
stinger at the back end, but it is just a
tool for the male to hold onto the
female while mating. They live by
scavenging dead insects, sometimes
even stealing them from spiders' webs.
They also eat live greenflies (aphids)
providing a valuable ecosystem service
to gardeners and farmers.

Niall Benvie

The Iron Gate
SERBIA/KAZAN GORGE, DJERDAP
NATIONAL PARK

The 100 km (62 mi.) long Djerdap
Canyon, on the border between
Serbia and Romania, has been carved
through the limestone rock by the
mighty Danube River: Europe's second
longest river, after the Volga in Russia.
On the Romanian side there is the Iron
Gate Natural Park. The 636 sq. km
(24 sq. mi.) Djerdap park is a candidate
for Unesco World Heritage Area status
and is included in the Southern
Carpathian Wilderness project, a
cross-border initiative proposed by
WWF International to link wild lands in
Romania and Serbia. This would create
a protected wilderness area of about
10,000 sq. km (3,860 sq. mi.),
6,200 sq. km (2,400 sq. mi.) of which
is already under different forms of
protection.

Ruben Smit

"Eighty-eight per cent of Europeans would like nature conservation to have the same influence on political decision making as economic issues.

Eurobarometer survey, 2005

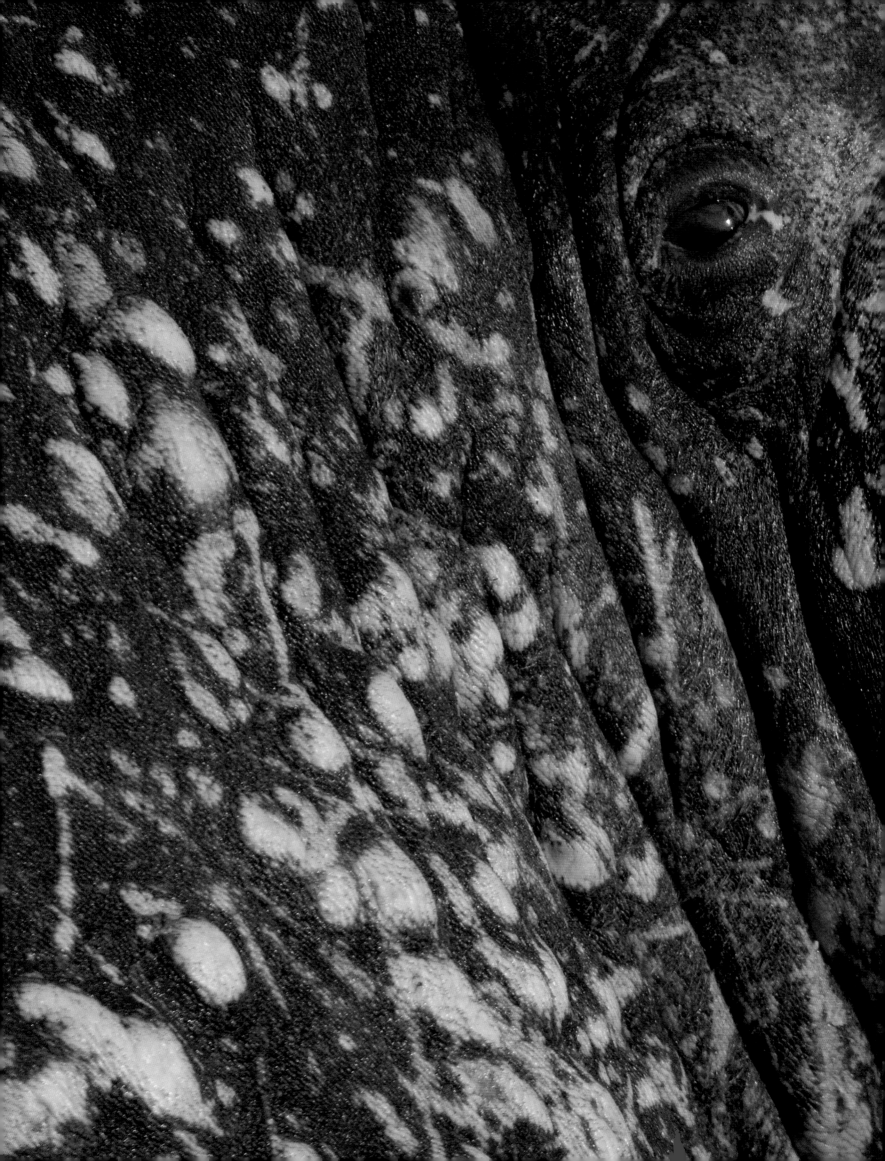

Walrus
Odobenus rosmarus
NORWAY/SVALBARD

After centuries of merciless exploitation that made it extinct throughout its entire European range, the walrus is back on stage. It feeds almost exclusively on small sand mussels, which it finds on the ocean floor with help from its stiff whiskers. A walrus can eat 8,000 mussels a day.

Mireille de la Lez

RIGHT
Red squirrel
Sciurus vulgaris
UNITED KINGDOM/INVERNESS-SHIRE, SCOTLAND

One of our most beloved and most widely recognized wild animals, the red squirrel is present over much of Europe. But in the British Isles and northern Italy it has come under threat from an introduced species from North America, the grey squirrel. Wherever the grey squirrel becomes established, the red quickly disappears. In recent years efforts have been made to control the spread of grey squirrels in the United Kingdom, but this is costly, time-consuming, and controversial. The introduction, deliberate or accidental, of alien, invasive species is one of the biggest threats to the variety of life worldwide.

Peter Cairns

Gannet face
Sula bassana
IRELAND/SALTEE ISLANDS, WEXFORD

The gannet is the ultimate flying machine, spending most of its life at sea. Gannets breed in colonies on Europe's rocky Atlantic coasts, often tens of thousands of them together. The Saltees are the most famous bird sanctuary in Ireland, with huge colonies of breeding seabirds—gannets, puffins, razorbills, fulmars, kittiwakes, and guillemots. The islands are privately owned but are open to visitors during the day.

Pål Hermansen

Common crane feathers
Grus grus
SWEDEN/LAKE HORNBORGA, VÄSTERGÖTLAND

Historically shot for food or because it was seen as a pest, the common crane is now a protected species in most of Europe. And their numbers are rising everywhere. At Lake Hornborga, they gather in spring, along with 150,000 crane-watchers, to perform their spectacular mating dance.

Stefano Unterthiner

Sunrise in the Julian Alps
SLOVENIA/MOUNT KRIZ, TRIGLAV NATIONAL PARK

Triglav is *the* national park in Slovenia. Mount Triglav, at 2,864 m [9,396 ft.], is the national symbol of Slovenia and is featured on the country's flag and coat of arms. Our natural heritage is a powerful part of our identity.

Daniel Zupanc

Greater flamingo
Phoenicopterus roseus
FRANCE/CAMARGUE REGIONAL NATURE PARK, PROVENCE

Flamingos get their pink colouration from the carotenoids in the small crustaceans that they sieve from the water with their strange-looking bills. In Europe, the flamingo breeds in Spain, Portugal, Italy, France, Cyprus, Turkey, Greece, and Albania. When the 820 sq. km [317 sq. mi.] bird reserve in the Camargue was created in 1972, thanks to the visionary conservationist Dr. Luc Hoffmann, one of the founders of the WWF International, it was a landmark victory for nature conservation in Europe.

Theo Allofs

The alpine ibex is back again in Chamonix
Capra ibex
FRANCE/LAC BLANC, CHAMONIX, HAUTES-ALPES

Mont Blanc, on the right, is the highest peak in the Alps at 4,810 m [15,781 ft.]. The ibex, on the left, has returned to this region after a century of absence and is once again a common sight, thanks to a series of reintroductions and strict hunting bans. It is amazing how empty a place suddenly feels when a species disappears—and how happy everyone is when it comes back. Losing a species leaves a hole in our souls.

Frank Krahmer

Unreplaceable

Wildlife in Europe is coming back because a number of smart decisions have been made. Natura 2000 is one, the biggest network of protected areas on Earth. And some places in Europe are even being "rewilded". So there is good reason to celebrate. But at the same time, we are still losing species and habitats because of other decisions that were not quite as bright. We are losing some of the variety of life itself—our biodiversity—mainly because of wasteful farming, forestry, and fishery practices that often are paid for with taxpayers' money. Wasting this irreplaceable variety of life is like sawing off the branch we are sitting on. It is simply not very clever. Slowly, steps are being taken in the right direction on the political level, but a lot more needs be done by all of us ourselves.

In a recent Gallup poll, 80 per cent of United Kingdom residents thought that biodiversity was a kind of washing detergent. Okay, biodiversity may be an academic term, but its real meaning is far more important to every one of us. We are talking about nothing less than the sheer variety of life on Earth: of every species, from the huge whales to invisibly small, microscopic bacteria. It also means the variety of all the different habitats where they, and we, live. At the very centre, biodiversity is about genetic variety, the basic building blocks of everything alive, the engine of evolution.

This variety of life is all around us, in the farmlands and forests of the countryside, as well as in cities and everywhere in between. We are a part of it ourselves, and we have a major influence on it.

From a selfish human perspective, it is also about what we need and have always needed to survive, since the beginning of time. It is about what we have for lunch and what we use to make our medicines and clothes. It is about what we wash ourselves in, what we drink, and what we cook our food with. And about what we build our homes with and how we heat them.

When we lose biodiversity, we lose part of ourselves, a part of our cultural identity and home, and a part of our future. We lose emotional values, we lose mental and physical well-being; and on a more spiritual level, we lose inspiration.

Unreplaceable

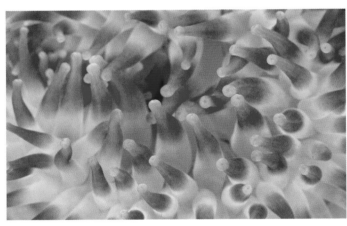

UPPER LEFT
Dahlia anemone
Urticina felina
NORWAY/SALTSTRAUMEN, BODØ
A flowerlike dahlia anemone folding its tentacles around its prey. Despite being one of the biggest fishing nations in the world, Norway's Atlantic-coast waters are remarkably rich in life. In many areas one can still talk about the living sea, with an abundance of fish and marine wildlife. This is in stark contrast to many other European marine areas, where overfishing has been the norm for decades. Norway shows that sustainable fishing is possible and that it can be achieved everywhere.
Magnus Lundgren

LOWER LEFT
Goat's beard or salsify
Tragopogon sinuatus
GREECE/PRINA, CRETE
The goat's beard is a common roadside plant across the Mediterranean. This one is from the remarkably wild island of Crete, a natural pearl of the Mediterranean and a true wildflower haven, with several plant species that exist nowhere else in the world.
Peter Lilja

RIGHT
Avocet with newly hatched chick
Recurvirostra avocetta
THE NETHERLANDS/TEXEL, FRISIAN ISLANDS, NORTH HOLLAND
A blue-legged avocet chick taking cover behind its mother's similarly blue legs. 70 per cent of Texel's economy is related to tourism. The beaches here are a major draw but so is Texel's amazing birdlife—the island is an El Dorado for wildfowl. Birds are very sensitive to environmental change and are excellent indicators of biological health.
Jari Peltomäki

PAGES 190-191
Waders in the Wadden Sea
GERMANY/WADDEN SEA NATIONAL PARK, SCHLESWIG-HOLSTEIN
Oystercatchers, plovers, and sandpipers at dawn in Hallig Hooge. Although much of Europe has been transformed by man, the Wadden Sea has largely escaped our intervention. We have built dykes and embankments on the coast for centuries, but the vast tidal mudflats and wetlands here remain one of Europe's primordial landscapes. It is now a Unesco World Heritage Area.
László Novák

The European Union and the Council of Europe have promised to stop further loss of biodiversity in Europe by 2010. That means right now. And it is a goal that has not quite been met. Not yet. But the target has actually come closer. Most freshwater ecosystems are in much better shape today than a couple of decades ago. The water is cleaner. Most bird species are stable in numbers or even increasing. Freshwater fish species and many larger mammals are coming back. But at the same time, a number of species are disappearing, some of them rapidly. In most cases it's due to how we have chosen to use our natural resources. Whereas wildlife in earlier decades disappeared because of persecution, overhunting, collecting, and poisoning by industrial pollutants, the main threats today come from how we use the land and the sea.

Every kind of landscape has its connected wildlife and plant life. One farming method gives one kind of wildlife. Another farming method gives another. And when you turn the farmland into industrial areas or cities, wildlife changes again.

Most old-growth forest species are having a hard time, as are species connected to old-style farming and grazing landscapes, since all these habitats are disappearing at an alarming rate, and with them the variety of life that used to live there. They are being replaced by monotonous industrial forests, exotic species, monocultures, industrial farming, and sprawling town- and cityscapes. And biodiversity changes accordingly. Some species disappear, others profit from it. We simply get the variety of life that we deserve.

Too bad for us all that these new landscapes are much poorer in wildlife and variety. Today about half of all the species and about two-thirds of all the habitats in the European Union are in what the scientists call "unfavourable conservation status". Fifteen per cent of the mammal species and 13 per cent of the bird species within the EU are listed as "threatened with extinction". Sixty-five per cent of Europe's habitats and more than half its remaining species are under threat. So there are several reasons to really worry and not just celebrate. There are many losers when it comes to biodiversity in Europe.

Almost all large insects are being killed by pesticides, grazing landscapes are disappearing, and they need the rotten wood of trees older than industrial forestry usually allows. This is bad news also for all species that prey on the large insects—rollers, woodpeckers, shrikes, lesser kestrels, red-footed falcons, and many different lizards.

Several of our classic farmland birds are decreasing because there are fewer insects to eat in the chemically treated fields of industrial farming. Birds like starlings, skylarks, lapwings, and swallows are decreasing rapidly.

Non-commercial plant species of all kinds are disappearing from the farmlands. Sometimes actively helped by plant-killing poisons and herbicides.

PAGES 198–199
The Rapa River delta
SWEDEN/SAREK NATIONAL PARK, LAPLAND
Sarek was one of the first national parks in Europe, established in 1909 and now part of the Laponia UNESCO World Heritage Area. A wilderness, but also an ancient cultural landscape for the Sami who have tended their reindeer flocks here for centuries. Every hill and every bend of the Rapa River has a Sami name. The oldest settlements found in Lapland date back around 6,000 years and were home to the ancestors of the modern-day Sami, who survived by hunting elk and reindeer, and fishing for salmon.
Peter Cairns

LEFT
Dusky grouper
Epinephelus marginatus
FRANCE/LAVEZZI ISLANDS MARINE RESERVE, CORSICA
A predator of fish, crabs, and octopus, the impressive 60 kg (132 lb.) dusky grouper was once a common species in the Mediterranean Sea; but like almost all major fish species of the region, it is now absent from huge areas. Only in marine reserves like this can groupers of any size be found today. This shows again the immense importance of marine reserves and the vital need to create many more of them.
Linda Pitkin

RIGHT
Black stork catching a fish fry
Ciconia nigra
GERMANY/ELBE RIVER BIOSPHERE RESERVE, NIEDERSACHSEN
The black stork is a bird of mythic status across its range. Historically, it was seen as a bad omen, in contrast to its white cousin. It is a shy creature of forest and mountain that breeds from Spain in the west to China in the east, with Poland having the largest European population, around 3,000 pairs. Most black storks spend their winters in Africa, but some, like the Spanish birds, enjoy the kind winters of the Mediterranean.
Dieter Damschen

Delicate species are killed or lose in competition with a limited number of tougher species, when meadows are enriched with chemical fertilizers. We get dandelions instead of 100 other species.

Lichens, fungi, and many plants need old-growth forest and dead wood to survive.

All species connected to the steppe habitats are in danger, since the steppes are being converted into farmland and/or fenced up into small parcels.

Most marine fish species of commercial value are heavily overfished, and almost all creatures living on the sea floor are damaged by intensive bottom trawling in huge areas.

RIGHT
European rainforest
GEORGIA/MTIRALA NATIONAL PARK
An ancient beech tree in a lush temperate rainforest close to the Black Sea. Mtirala is Georgia's newest national park and occupies the area of the ancient state of Colchis, known from the earliest Greek myths. Rhododendron species grow in many European mountain ranges, but usually as low, creeping bushes. In Georgia, there are some low-altitude rhododendron species, at times forming large stands or thickets, which flower spectacularly in May.
Georg Popp

PAGES 204–205
Greek land tortoise
Testudo graeca
GREECE/LAKE KERKINI, MACEDONIA
Land tortoises benefit from the dry lands that are created by intensive grazing by sheep, goats, and donkeys. Originally, the wild grazers here were most probably mouflon sheep, ibex, red deer, and tarpan horses. The tortoises are now disappearing because the sheep are, too. All across Europe, vast lands far away from major towns and cities are being abandoned, and ancient grazing areas turn into bush and forest. This paves the way for the reintroduction of native wild grazing species, providing an opportunity for the rewilding of quite large areas.
Staffan Widstrand

Too-intensive farming is causing severe damage to the web of life, with huge tracts used for growing just a few crop species. Heavy use of pesticides and chemical fertilizers knocks out a lot of the insects and wildflowers, and overfertilizes lakes, rivers, streams, and in the case of the Baltic, even the sea. Since the 1940s, around 98 per cent of the flower-rich meadows in the United Kingdom have disappeared, taking with them many butterflies. Wetland butterflies have decreased there by 90 per cent and grassland species by 30 per cent since 1970. Cattle today are often doomed to a life indoors, creating huge effluent problems and a loss of insect and plant life in the fields where the cattle used to graze. And many of these fields are now turning into scrub and bushland, with an impact on many bird species.

Most of the Mediterranean forests are gone. Only about 30 per cent remain today, and many of them suffer from exploitation and arson. The proportion of remaining old-growth forests in the European Union is estimated at less than 1 per cent of the total forested area. Climate change is an added threat to a number of species, especially insects, amphibians, and alpine plants that live in isolated areas.

But right now, by far the worst disaster of all is unfolding in our seas. The mismanagement of European fisheries is one of the biggest environmental scandals of our times. More than 80 per cent of the commercial fish stocks in Europe are being carelessly overfished. Almost half of the total fish stocks are not within safe levels, with a third at risk of being beyond recovery. The Mediterranean, one of the Earth's "Biodiversity Hotspots", is becoming an empty sea. Very valuable fish species like the bluefin tuna, once incredibly common, have now all but vanished. The Baltic Sea is severely damaged by overfishing, as is the Black Sea.

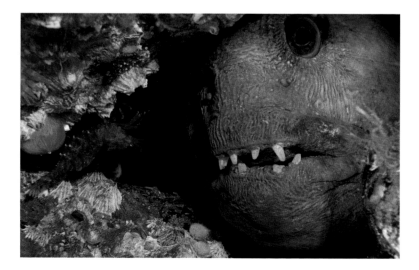

LOWER LEFT

Atlantic wolf fish and shrimp
Anarhichas lupus and *Lebbeus polaris*
NORWAY/SALTSTRAUMEN, BODØ
The wrinkled blue-grey wolf fish is a highly prized delicacy that grows very slowly and can live for decades. Many wolf fish accommodate this shrimp species in their lair. It eats the scraps from the wolf fish's favourite meal: sea urchins.
Magnus Lundgren

UPPER RIGHT

Eleonora's falcon
Falco eleonorae
GREECE/ANTIKYTHERA, ALONISSOS MARINE RESERVE
This was the first bird species ever to be protected in Europe, by a royal decree from Queen Eleonora of Arborea in Sardinia, Italy, way back in 1395. A specialist in catching small migrating birds, this rare falcon breeds late in September to benefit from the huge number of birds on their way to Africa for the winter. Thanks to early pioneering conservation policy, the Eleonora's falcon still breeds in good numbers on remote islands in the Mediterranean, but nowhere else in the world.
Stefano Unterthiner

LOWER RIGHT

Long-nosed viper
Vipera ammodytes
SERBIA/DJERDAP NATIONAL PARK
Vipers are perhaps not everyone's favourite creatures. But they perform a very important ecosystem service to man: farmland pest control, since they specialise in eating the very rodents that would otherwise eat farmers' crops.
Ruben Smit

PAGE 208

Great grey owl
Strix nebulosa
FINLAND/OULU, POHJOIS-POHJANMAA
The great grey owl is one of the most charismatic birds on the planet, but because of prejudice and superstition was often shot in earlier times. This magnificent rodent specialist is now reclaiming lost territories and slowly expanding its range. It is a very confident bird, often active during daylight hours and therefore more easily spotted than many other owls.
Sven Začek

PAGE 209

Male staghorn beetle
Lucanus cervus
UNITED KINGDOM/SUFFOLK, ENGLAND
This is the biggest beetle in Europe, often growing to the size of the width of a man's hand. Uncommon, and mostly seen at dusk during their swarming period in June and July, the staghorn prefers old woodland with ancient oaks, and gardens with old fruit trees.
Niall Benvie

"Same smells, same sounds, same tastes—who wants boring monotony? Variety instead adds life to our lives."

Peter Cairns, Wild Wonders of Europe photographer, Scotland

These are all the results of thousands of not-so-wise decisions. Big and small, made over the decades by single individuals and by companies, as well as by major political assemblies. The good news is that these same entities can start making better decisions: in favour of the variety of life, in favour of biodiversity, instead of against it. And this has already begun to happen on many levels.

It started 30 years ago, back in 1979, when the member states of the Council of Europe and a few North African countries signed a legally binding document called the Bern Convention. Its aim was to save the wild fauna and flora of Europe together with their natural habitats and to promote European cooperation in that field. At the same time, the European Union created the legally binding Birds Directive, aimed at protecting the birdlife within the member states. This was followed in 1992 by more binding legislation, the Habitats Directive, aimed at protecting biodiversity through a pan-European system of protected areas. This in turn led to the creation of the Natura 2000 protected areas network and a parallel Emerald Network, which is basically the same thing but includes non-European Union countries in Europe. Out of 48 European countries, the only ones who have not yet signed the Bern Convention are Belarus, San Marino, and the Vatican.

Today, in 2010, Natura 2000 includes about 25,000 sites in all of the 27 member states, representing almost 20 per cent of European Union territory.

PAGES 210–211
Loggerhead turtle and pilot fish
Caretta caretta and *Naucrates ductor*
PORTUGAL/PICO, AZORES
A loggerhead turtle cruising the great blue ocean, far away from land, followed by a school of pilot fish. The pilot fish help the turtle by ridding it of parasites. Pilot fish also follow sharks, often the same individual for years. Sometimes they even follow ships, hence their popular name. In 2008 no fewer than 55,000 sea turtles were accidentally caught by European vessels on longlines designed for catching swordfish.
Magnus Lundgren

LEFT
Northern bluefin tuna
Thunnus thynnus
MALTA
Bluefin tuna are large, long-lived fish that normally live in huge schools. They can reach 30 years of age and the heaviest tuna recorded to date weighed 680 kg (1,500 lbs.). The Atlantic population of bluefin tuna has decreased by more than 90 per cent since the 1970s. In recent years, the European fishing fleet has been taking 60,000 tonnes (66,138 tons) of tuna annually, beyond the estimated sustainable harvest of 7,500 tonnes (8,267 tons). This is a fish that we should all immediately refrain from eating! The rarity of the tuna required that this photo be taken in captivity, at a breeding facility.
Solvin Zankl

RIGHT
Sea holly
Eryngium dilatatum
PORTUGAL/COSTA VICENTINA, SOUTHWEST COAST NATURAL PARK, ALENTEJO
Sea hollies are among the few plants that have learned to survive the extreme conditions of European ocean and beaches, exposed to salt water, wind, scorching sun, and a lack of fresh water. The Costa Vicentina is a Unesco World Heritage Area.
Luis Quinta

It covers some 700,000 sq. km (270,000 sq. mi.) of land surface and 150,000 sq. km (58,000 sq. mi.) of the sea. That is an area about three times the size of Britain, twice the size of Poland, or twenty times the size of the Netherlands. This is extremely important to the future of Europe's natural heritage and is a major milestone in international nature conservation. Natura 2000 is specifically aimed at stopping the destruction of high-biodiversity habitats, and it is one of the main factors behind the European wildlife comeback over the last 20 years.

But it doesn't stop there: the European Parliament, as recently as 2009, backed a report calling for further protection of Europe's wilderness. The report also calls for more European Union funding to protect existing sites and to rewild some areas that are still being used by humans. Rewilding is a relatively new concept that has just begun to be tried out in various parts of Europe. The idea is that unused, abandoned, or waste lands can be brought back to a wild or wilder state by actively letting the original species from that habitat reshape the landscape, especially large browsing and grazing mammals.

RIGHT
Scorpionfish couple on an artificial reef
Scorpaena porcus
MONACO/LARVOTTO MARINE RESERVE
Scorpionfish have two forms of protection, poisonous spines on the back and camouflage. Do you see the second one on the left? Monaco and its head of state, Prince Albert II, have taken a leading role in marine conservation. The Larvotto Reserve, created right outside the harbour of Monte Carlo, is one example. Albert has also been instrumental in creating the huge whale and dolphin reserve north of Corsica, the 90,000 sq. km (34,750 sq. mi.) Pelagos Sanctuary, and has been active in international diplomacy to save the magnificent bluefin tuna.
Franco Banfi

Oostvaardersplassen in the Netherlands is a European mini-Serengeti. Here koniks, horses of a race closely related to the now-extinct tarpan horse, together with Heck cattle, close relatives to the now-extinct wild aurochs, were introduced in 1983 and 1984 and simply let loose on a piece of wetland, some 7,000 ha (17,300 ac.) in size. In 1992, red deer were also introduced. And today, after 25 years of completely unmanaged grazing, the 2,500 animals have recreated something similar to what a European savannah must have looked like at the dawn of human history. The evidence shows that this savannah is capable of supporting numbers of large wild mammals per hectare similiar to those on many of the best African nature reserves.

Oostvaardersplassen is an exciting pioneer area and a tantalising example of what can be achieved. Could an even a larger "European Serengeti" be created somewhere? Well, steps towards rewilding are being undertaken in a number of places. Among them are Hortobágy, Hungary; Bieszczady, Poland; Peneda-Gerês, Portugal; and Alladale, Scotland. Several other areas are being considered for rewilding initiatives. By far the largest rewilded area, though, lies in the borderlands between Ukraine and Belarus. This happened more or less instantly after the nuclear disaster in Chernobyl in 1986 and the resulting fallout of radioactive particles. 371,000 people were forced to evacuate the area within a week because of dangerously high radiation levels, and since then this area of about 430 sq. km (170 sq. mi.) has been almost completely closed to people. This has led to uniquely high populations of all kinds of wildlife, especially moose, wild boars, roe deer, wolves, and red deer. The whole area is now classified as a nature reserve in both countries. Przewalski's horse, the world's last remaining truly wild horse species, has been introduced on the Ukrainian side, and bison on the Belarussian side.

The fact that many millions of Europeans are leaving the countryside for city life is also adding to this process. It has been estimated that by 2035 some 700,000 sq. km (270,000 sq. mi.) of former farm and grazing land could be abandoned, especially in the more remote and mountainous areas of southern and eastern Europe. Today, land abandonment often leads to a severe loss of species, as old farmlands and grazing lands turn into forest or scrub. Bringing back the wild grazing species could change that. Many of these abandoned areas might be candidates for future rewilding initiatives.

PAGES 216–217
Koniks
Equus ferus caballus
THE NETHERLANDS/
OOSTVAARDERSPLASSEN
NATURE RESERVE

The konik is considered the nearest direct descendant of the tarpan, which lived in Europe until it was hunted to extinction in 1887. A group of koniks that were released in Oostvaardersplassen 30 years ago, have, together with Heck cattle (genetic reconstructions of the aurochs, extinct in 1627) and red deer, re-created a European savannah of a kind not seen for over a thousand years. This has showed that a "European Serengeti" could carry at least as many wild animals per hectare as the best wildlife areas in Africa. Possibly even more. Where will the next real European savannah be re-created?

Mark Hamblin

RIGHT
Female goldenrod crab spider on orchid
Misumena vatia and *Dactylorhiza sp.*
SAN MARINO

Crab spiders are well adapted to their role as stealth hunters. They don't make cobwebs, but instead take position in colourful flowers, where they wait for nectar-loving insects to land. This species can even change colour depending on which flower it uses.

Florian Möllers

PAGES 220–221
Iberian lynx
Lynx pardinus
SPAIN/SIERRA DE ANDÚJAR, JAÉN

The Iberian lynx is the most endangered cat species in the world, native to Spain and Portugal, but today only present in two small populations, one in the Coto Doñana and a larger group in the Sierra Morena area, both in southern Spain. If it disappears, it will be the first extinction of a cat species in the world since the sabre-toothed tiger 10,000 years ago. The main reason for its decline is a major crash in the population of its main prey, the European rabbit, which has been struck by introduced diseases like myxomatosis and rabbit haemorrhagic disease. Habitat loss, trophy hunting, road construction, poisoning, and snaring are all additional threats to the lynx. The present estimate of 200 adult individuals is down from an estimated 100,000 in 1900, but up from 135 just a few years ago. In 2010, the first lynxes bred in captivity will be reintroduced to suitable areas.

Pete Oxford

Nature has always provided us with a lot of life-supporting services for free. Pure water; fertile soils; clean air; insects that pollinate our crops; forested mountain slopes that protect against avalanches, landslides, and floods; wetlands that are crucial for water purification; peat bogs that are important for carbon storage; and populations of fish, game, mushrooms, and berries that we can harvest. This is all delivered by nature and its variety of life.

If we don't ruin it, that is.

Slowly, our society is beginning to realise that these so-called ecosystem services are more valuable to us than we might have thought before.

At last we seem to be learning that damaged ecosystems become very expensive for society, and that conservation can be profitable and make economic sense. In order to avoid further waste, some say that the time has come to put a price tag on those services from nature. For example, farmers and landowners can be paid for keeping or restoring the variety of life on their property, using public money for a key public good: ecosystem services rather than subsidies for industrial farming and agrobusiness.

Mouflon sheep
Ovis musimon
FRANCE/HAUT-LANGUEDOC REGIONAL NATURE PARK, MASSIF CENTRAL
The mouflon is the ancestor of most of the world's domestic sheep, probably first domesticated in Anatolia in Turkey. The original range of wild mouflon in Europe is disputed. Today, ancient populations of mouflon exist on Sardinia and Cyprus, as well as in Turkey, but some scientists believe that mouflon were brought to the Mediterranean islands by early man, thousands of years ago. Mouflon have since been introduced into several other areas in Europe, such as here in Languedoc, primarily for hunting.
Ingo Arndt

Bohemian hills
CZECH REPUBLIC/ČESKÉ SVÝCARSKO NATIONAL PARK, BOHEMIA
The view from Vilhelmina Vyhlidk, looking out over the mixed pine and broadleaf forests that cover most of this karstic limestone landscape, is familiar to the visitor to this 79 sq. km (30 sq. mi.) national park, created in 2000. Its name means "The Czech Switzerland" and the park is connected with the Sächsische Schweiz National Park on the German side of the border.
José B. Ruiz

At the end of the day, saving the variety of life is up to all of us. Our attitudes and actions, what we choose to think, and what we choose to do. The decisions we make have immediate consequences— good or bad. We have the power to wipe out wildness completely, but also to let it come back and flourish. Some of that action needs to be taken on a political level, but our own daily individual actions and decisions are what really count. How we vote with our own wallets. What we choose to buy and what we choose not to buy. What we do and what we don't do. Almost everything can still be fixed.

Until something irreplaceable is finally lost.

RIGHT
Monk seals flirt
Monachus monachus
PORTUGAL/DESERTAS ISLANDS, MADEIRA NATURAL PARK

This unique image of a pair of monk seals in mating behaviour gives some hope for the future of this, the world's most endangered seal species. Once common all across the Mediterranean, depicted on Greek coins from 500 BCE and seen as an omen of good fortune, there are today fewer than 600 individuals alive. Depleted by commercial harvesting in Roman times, and persecuted by fishermen since, here in the Desertas, their numbers were down to just 6 individuals in 1988. Now there are at least 36. An improvement but still a painfully vulnerable small group.
Nuno Sá

PAGES 226-227
Dab
Limanda limanda
NORWAY/LOFOTEN, NORDLAND

The dab is a common demersal Atlantic fish species, which spends its time patiently waiting for something edible to fall down to the ocean floor. In order to save the future of European fish stocks, a number of actions are urgently needed: the fishing quota system needs to be revised completely, marine research needs to be disconnected from the fishing industry, better labels of origin need to be developed, better enforced fishing laws and stricter penalties need to be applied, the dumping of bycatch needs to be banned, bottom trawling needs to be banned, and the creation of many more marine reserves is needed, especially in the most productive spawning areas.
Magnus Lundgren

"The unique European Natura 2000 system is the largest network of protected areas in the world. It is something to be very proud of and happy about!

Magnus Sylvén, conservation consultant, Switzerland

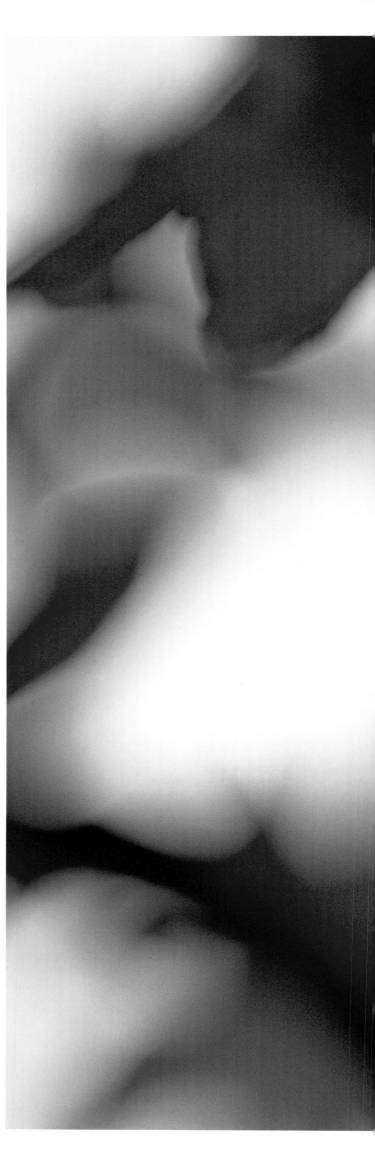

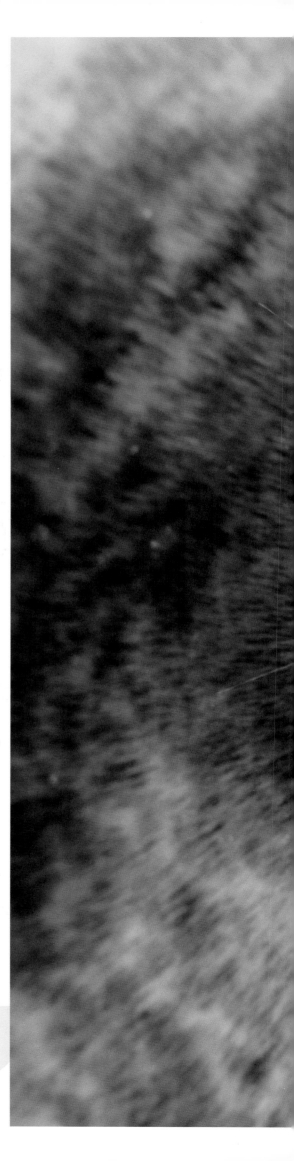

Unreplaceable

"There is no reason why Europe could not
have major wildlife areas with big numbers of
large wild animals.

Johan van de Gronden, CEO, WWF Netherlands

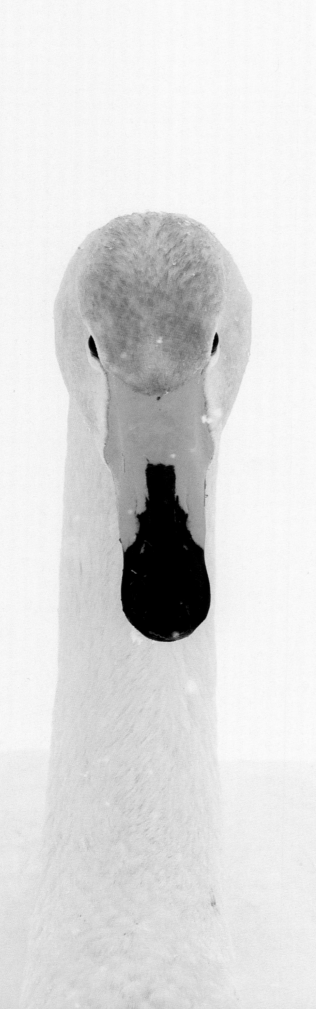

Unheard of

The real heroes of the Wild Wonders of Europe initiative are the 68 top photographers who carried out no fewer than 125 assignments among them, to every one of Europe's 48 countries. That is a world first. The first time the crown jewels of the natural heritage of an entire continent have been photographed in one single campaign. The biggest-ever professional nature-photography initiative. Most of these artists were sent to a country other than their own, in order to come back with a fresh take on their subjects. Well, they sure did. Fourteen months of field work and 170,000 pictures later, they have returned from deserts, forests, and mountain tops; from tundras, steppes, and the deep seas. One of the results is the selection of over 200 images for this book. Here are some of the stories behind the making of these images: unheard tales from the bush.

On my arrival in Dovrefjell in Norway the thermometer shows −27°C (−17°F). The snow squeaks under my snowshoes, and my breath freezes in my nose, on my eyelashes, and in my hair. My heavy sledge weighs 50 kg (110 lbs.), loaded with winter camping and photo equipment. The landscape is very quiet, there is no wind; and when a mist slowly comes up, I hear some barking foxes in the distance. For the rest of the day, it is the silence of the snow that rules. A few days later, I manage to find and approach a clan of 19 musk oxen. I stay for hours and hours in their presence without seeing them even move an ear—they are just like snowed-over rocks; and I have to try to adapt to their rhythm . . . no movement at all. But suddenly, out of nowhere, a storm rolls in over the snow-covered mountain tundra; and it starts looking like a wild sea, with waves of snow beating against my body. The musk oxen liven up and start moving around, maybe to find a more protected position. I keep shooting pictures in the incredible light as the sun shines through the dark clouds during the storm. The sound of the hard wind is fantastic. At the end, not only are my fingers frozen stiff, but what is worse, my camera and my 70–200 mm lens are frozen. During the last couple of hours the temperature has fluctuated from −10°C to 2°C to −20°C (14°F to 36°F to −4°F), which caused condensation, the winter photographer's worst enemy. Time to head back to the log cabin to thaw out the equipment. And the photographer.

Vincent Munier, France

Vincent Munier in the huge landscape.
NORWAY/DOVREFJELL-SUNNDALSFJELLA NATIONAL PARK

It takes both guts and field craft to ski straight out into the wilderness and then camp there for weeks in −25°C (−13°F) temperatures, while trying to find and approach the musk oxen without scaring them away or provoking them to attack.
Vincent Munier

Musk ox faces
Ovibos moschatus
NORWAY/DOVREFJELL-SUNNDALSFJELLA NATIONAL PARK

The musk ox was one of the typical tundra species that lived together with reindeer and mammoth on the tundras of the Ice Age. The mammoth and musk ox were easy to hunt and were quickly exterminated by early man. The mammoth disappeared forever, but the musk ox survived in the extreme north of Greenland and Canada and has since been reintroduced to several sites in Scandinavia, Greenland, and Russia, as here in Dovrefjell.
Vincent Munier

Wild reindeer bulls fighting
Rangifer tarandus
NORWAY/FOROLLHOGNA NATIONAL PARK, HEDMARK/SØR-TRØNDELAG

Each autumn, reindeer bulls compete for the right to mate with as many females as they can keep in their harem. Fights can be explosive and occasionally deadly, with only the strongest bulls winning the ultimate prize. For the losers, more practice is needed; and for some, their time will come.
Vincent Munier

One evening, about 20 km (12 mi.) from the Chernye Zemli Biosphere Reserve in Kalmykia, Russia, our car comes up over a low sandy ridge. A green valley opens before us and our eyes immediately fall on a herd of about 100 saiga antelope, racing at top speed over the steppe, with three motorbikes in close pursuit. Poachers! They notice our car atop the hill and, without slowing down, turn in different directions, like market thieves. I take several images of the saiga antelope racing along the road. I see through the 300 mm lens that they are males with horns. The horns are what the poachers are after. By cell phone, we contact the ranger patrol, who are in ambush on the northern border of the reserve buffer zone, and the rangers arrive about ten minutes later. We follow the tracks in the grass. One set of tracks leads to a shepherd's house, outside of which a sport bike is standing, with the engine still hot.

A dusty young man from the Caucasus heartily milks a cow and says, "I just herded the cows in from the steppe, that's why the bike is still warm". The tracks of another motorcycle lead to the gates of another shepherd's house, and the third emerges onto the paved road. Hopefully, our presence at the scene was at least enough to win the chased saiga males enough time to get back to the safety of the nature reserve. There, over an area of about 1,200 sq. km (460 sq. mi.), the saiga have created an enormous kindergarten, with at least 10,000 adults and their young—most of the remaining saiga in Europe.

We drive with headlights turned off, into the midst of the saiga herd under the bright stars of the southern skies. Well before dawn we hear the baaing of thousands of saiga calves and their mothers. The steppe is coming alive.

Through an opening in my hide, I can see hundreds of dim silhouettes

Saiga antelope male
Saiga tatarica
RUSSIA/CHERNYE ZEMLI BIOSPHERE RESERVE, KALMYKIA

The trunklike nose of the saiga is a bizarre adaptation to dust—an ingenious filter system. Even if there is no wind, a herd of saiga running at 70 km/hr. (43 mph), kicks up plenty of dust on the dry steppe. The saiga's nose also helps their calls resonate across the vast open grasslands they call home. Chernye Zemli Biosphere Reserve is a Unesco site, created in 1990 to protect the saiga antelope, whose numbers have crashed due to poaching and desertification caused by overgrazing by domestic livestock.

Igor Shpilenok

UPPER RIGHT
Steppe eagles
Aquila nipalensis
RUSSIA/CHERNYE ZEMLI BIOSPHERE RESERVE, KALMYKIA

The steppe eagle is a common raptor on the open steppes of Asia and easternmost Europe. It lives mainly on small rodents, especially ground squirrels or susliks, but will take prey as large as a young saiga. It also happily eats lizards and insects like swarming locust grasshoppers. Perhaps surprisingly, they nest right on the ground.

Igor Shpilenok

LOWER RIGHT
Saiga antelope calves
Saiga tatarica
RUSSIA/CHERNYE ZEMLI BIOSPHERE RESERVE, KALMYKIA

By 1950, the saiga antelope had a population of around 800,000 living on the remaining wild steppes on the banks of the Volga River and near the Caspian Sea. After the fall of the Soviet Union, poaching became rampant as saiga horn is highly prized in traditional Chinese medicine. Today, the entire European saiga population is down to fewer than 18,000. Since poachers mostly target the males, male saiga now make up only 1 to 10 per cent of the population. Those numbers are too low to guarantee the species' survival, and the saiga antelope is moving quickly on the road towards extinction. This is one of today's major wildlife tragedies, and it's happening right here in modern Europe.

Igor Shpilenok

of the antelope passing by, followed by their calves. The place I picked for the hide couldn't have been better—the saiga are passing right beside it. Finally, in the early morning light, I can start taking pictures, wrapping my Nikon D3 in a sweater to muffle the sound from the shutter; but its noise still seems deafening. The animals fortunately don't seem to mind. During the day, a strong wind picks up the dry sand, covering my lenses and cameras. I put them away, closing all the openings in the tent; but the dust still manages to penetrate the tent, into my eyes, nose, and ears, and I feel grit between my teeth. It dawns upon me that for a whole month I hadn't had a decent bath, eaten a decent meal, or seen my family. But I hope what I do here might be helping to save the incredible saiga.

Igor Shpilenok, Russia

Demoiselle crane
Anthropoides virgo
RUSSIA/CHERNYE ZEMLI BIOSPHERE RESERVE, KALMYKIA
A characteristic bird of the great steppes of Asia, the demoiselle crane also breeds in small numbers in the steppe regions of easternmost Europe, in Russia and Ukraine. Their numbers are decreasing, mostly due to illegal hunting. In this 2,119 sq. km (818 sq. mi.) reserve, at least, they are reasonably safe.
Igor Shpilenok

"For the first time in history, eagles, goshawks, grouse, and gulls are worth more alive than dead.

Ole Martin Dahle, eagle-watching outfitter and former farmer, Flatanger, Norway"

On the wintry beach of the Trondheim fjord in Norway, my rock-steady dive buddy Patrik and I trudge our way through knee-deep snow in full scuba gear. It is heavy work. The snow gives way, and it feels a bit like walking in loose sand. The cold is biting; but by the time we reach the waterfront, I am breathing like a steam engine. I feel very hot in my bulky dry suit. All I want is to get into the cooling 5°C (41°F) water and finally reach zero gravity. I signal that we should go down, and we descend together into the black fjord on another search for the fairytale monster. Hopefully, the mysterious ghost shark will materialise. Normally, they live at depths between 200 and 1,000 m (660 and 3,280 ft.), but in some places in the Norwegian fjords they can be encountered in waters shallow enough for divers to see them. This is one of those places.

We follow the steeply plunging rock wall down as we search the area, and I feel my senses come alive. Two bulky stone crabs are fighting on a big rock. They look impressive, but we stay in focus and continue.

Then . . . nothing, a great nothingness again. The wall is steep and bare. We pass some sandy terraces and then out into black water again. Time to go deeper, hoping that something will be attracted to our lights. It is quiet and cold, an absolutely crisp and thrilling moment. The exhaust bubbles keep reminding me that I am actually diving. We reach our maximum depth and already it is time to start heading up. But right then, a pair of big, shining, green zombielike eyes can be seen far in the distance. All I can think is, "Mr Ghost Shark, you are late!" The shark swims gracefully along the wall, crossing one of the sandy terraces. It tilts its body forward and sweeps the sand with its nose, sensory systems searching for hidden food. The appearance of this strange, ancient shark is suddenly not strange at all. It is perfectly designed to do exactly that.

It is a magnificent animal, flying like an angel through the water, with its elegant long pectoral fins. After a few minutes the shark is attracted to my light and comes at me head on, only to turn away a centimetre or so from the front of my underwater housing. I think, "wow" and, "don't forget to come back" at the same time. Then it does it again and again and then another close shave.

The encounter goes on for a good six to seven minutes. We feel so fortunate, and with smiling faces rise slowly from the depths. A bit higher up, we have another encounter, possibly with the same individual; and when it swims out from the wall, I follow it. It feels surreal to leave the wall and enter the "sea without reference" in the blackness. I keep my eyes to the viewfinder and let my ears pay attention to depth change. I rely on this for a short while. We swim together for a brief but seemingly endless moment, until the

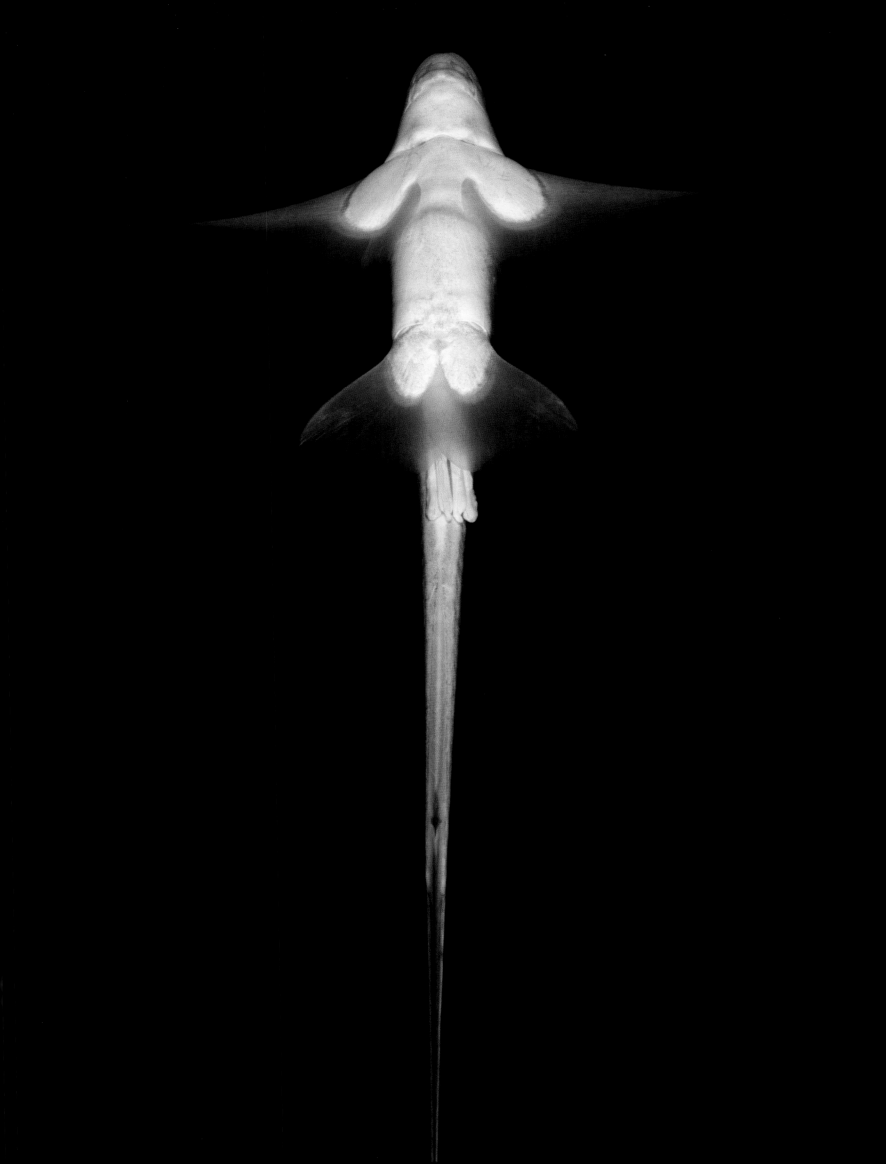

ghost shark gets a bit nervous and
decides to dive, fading away into
the deep.

Patrik's thumb is telling me firmly
that it is time to go up, and it takes
us around 30 minutes to reach the
surface. As we are getting out of the
swell, Patrik looks like it is Christmas
Eve, and I am of course also very,
very happy. To be out in the wild and
to experience intense wildlife
meetings as we just did deepens our
connection to Mother Nature. The
value of biodiversity becomes so
obvious and the importance of it
complete. To do this with a great
friend is a true memory for life.

Magnus Lundgren, Sweden

Black moray eel
Muraena augusti
PORTUGAL/FAIAL, THE AZORES

A black moray eel hiding in its favourite
crevice, waiting for nightfall, when it will
venture into the open sea to hunt.
Different moray eel species are present
across all the world's tropical and
subtropical seas.

Magnus Lundgren

PAGE 262
Plumose anemone
Metridium senile
NORWAY/LOFOTEN, NORDLAND

The intriguing mouth of a plumose
anemone, surrounded by its tentacles.
Anemones are predators that pacify
their small prey with poison from the
nettlelike stingers in those beautiful but
deadly tentacles.

Magnus Lundgren

PAGE 263
Nudibranch
Polycera quadrilineata
NORWAY/MØRE OG ROMSDAL

This colourful underwater snail, or
nudibranch, is one of the most common
of its kind in the North Atlantic. Nudibranches vary greatly in size, shape,
and colour and are sometimes referred
to as the "orchids of the sea". Scientists
estimate that they have only disovered
half of all nudibranch species existing
on our planet. The nudibranches are
hermaphrodites, which means that
each individual is both male and
female—a handy strategy, since every
fellow member of your species can then
be your lover.

Magnus Lundgren

"The joy of meeting wild, free beings is deeply rooted
within our genes. Be it a hedgehog or a squirrel, as well as
the soaring eagle."

Staffan Widstrand, Wild Wonders of Europe photographer, Sweden

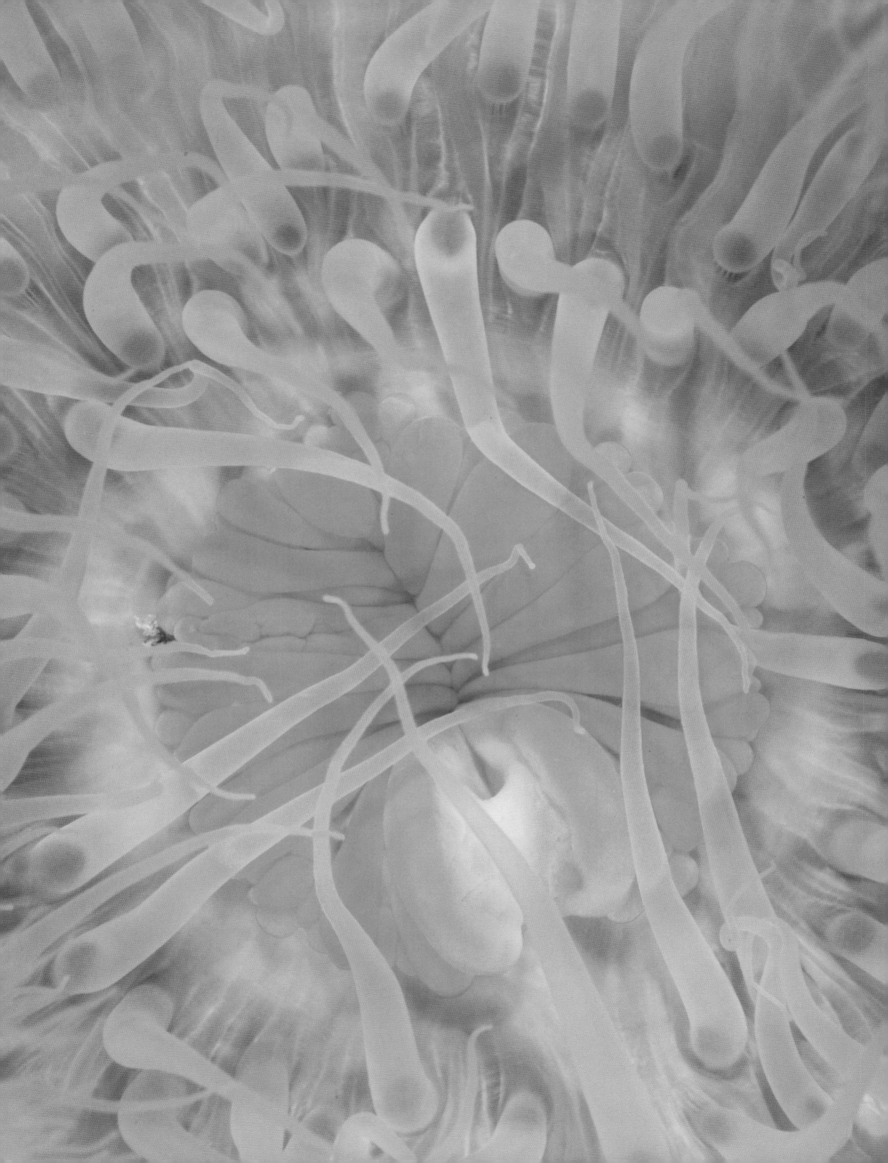

The orchids of Gargano are in most cases very beautiful, and we dig deep into our camera bags to try to portray them appropriately. The hard wind creates a lot of difficulties. Tripods and raincoats are set up as sails to soften the harsh light. Luckily, we are fans of maximum apertures, and we are able to use shutter speeds from 1/250 s upwards for orchids that are swinging happily in the storm.

We twine around sharp-cornered stones, thistles, and cowpies like fakirs—trying to picture the selected beauties in the best light possible. Bruised, scratched, sunburned, and full of black-and-blue marks, we agree that the road to success is paved with thorns. Many of them. There is real struggle and pain behind every picture that we manage.

So we decide to take a break—there are quite a few types of ice cream that we haven't tried yet, and the Illy espresso is really good. Driving the coastal road to Vieste, we suddenly discover a hillside with white and pink spots all over. We agree to take a closer look, and there—in between burned branches and Naples garlic—we find a few orchid species that are new to us. Within seconds we are highly motivated again. Ice cream can wait.

As always, you come across some of the best things in life where you wouldn't expect them. It is the same with orchids. We found some of the most beautiful specimens in overgrown landfills or in roadside ditches. The record 69 species of the Gargano region blossom at different times in the year, usually from March to September, although April and May are the best months to visit. On one of our last days here, we finally found a magnificent and huge lizard orchid [*Himantoglossum hircinum*] almost 1 m (3.3 ft.) in height, in between a small and winding mountain road, green scrub, and a steep rock face. With all the tricks that acrobatics, focal lengths, and apertures offer, we almost manage to cope with the situation photographically. In any case, we are happy—our first great lizard orchid.

Sandra Bartocha, Germany

RIGHT
Promontory orchid
Ophrys promontori
ITALY/MONTE GARGANO NATIONAL PARK, APULIA

The minute orchids from the incredibly diverse and decorative genus *Ophrys* are common during early spring in the Mediterranean countries. They prosper particularly where the land is periodically grazed. Monte Gargano holds more than 60 species of orchid and is thereby the European record holder for the number of orchid species found within a single area.
Sandra Bartocha

PAGES 266-267
Whooper swans
Cygnus cygnus
SWEDEN/LAKE TYSSLINGEN, NÄRKE

Previously killed and eaten over much of the north and therefore forced to breed only in the most remote areas, the whooper swan has made a huge comeback over the last 30 years, expanding its breeding range each year. Sweden had 30 pairs breeding in 1970, but more than 6,000 in 2007. The whooper's honking call is a sure sign of the northern spring, and the whooper swan is the national bird of Finland.
Stefano Unterthiner

It was a hard start walking the steep path up to the mountain refuge, Rifugio di Sella, in sleet and rain. With full photo equipment and camping gear for more than a week and suffering from both a cold and fever, it took nearly four hours to walk from 1,666 to 2,588 m (5,466 to 8,491 ft.) to reach our new home for the coming weeks: a single room for hikers passing by, with 14 beds without heating facilities. Primitive but luxurious at the same time, since the only alternative would have been a tent. From the early morning hours, we tried to follow different chamois on the steep terrain, climbing 800 to 1,000 m (2,625 to 3,280 ft.) up and down every day. It is hard work to try to keep up with these fast animals, but I felt my fitness improved considerably during these weeks of chamois and ibex photography.

After a few days it started snowing heavily, and it simply did not stop. We were still not especially worried when it reached 0.5 m (1.6 ft.). Not even when the first avalanches started sliding down the far away mountain sides. But one morning we woke up to 1.5 m (4.9 ft.) snow around the refuge, and it was still snowing heavily. I have never experienced anything like this before, and I am from a country where snow is commonplace.

Extreme weather conditions make life tough for animals living at high altitude, and it does not make things easy for the wildlife photographer either. But this is how nature photography is. One has to work together with nature.

The steep mountainsides, where we had initially been working, were now simply too dangerous after we noticed more avalanches closer to us. But when suddenly one avalanche broke through the little valley right between the two of us, while we were trying to approach two different animals just a few hundred metres from each other, we stopped for a minute and started to think about our own safety. We decided to get out of there. It took many hours to fight our way down the first few hundred metres, as the snow sometimes reached our armpits, even after treading it down.

Erlend Haarberg, Norway

RIGHT
Chamois
Rupricapra rupricapra
ITALY/GRAN PARADISO NATIONAL PARK, VALLE D'AOSTA

Once almost pushed over the edge of extinction because of overhunting, the chamois is now bouncing back all across the Alps, Pyrenees, Cantabria, and the Carpathian mountains. More sustainable hunting practices, better hunting laws, and stricter enforcement have brought this and a number of other iconic European wildlife species back from the brink of extinction. Nowadays you can often see chamois even in some of the most famous Alpine ski slopes.
Erlend Haarberg

PAGES 270-271
Alpine pink landscape
ITALY/GRAN PARADISO NATIONAL PARK, VALLE D'AOSTA

Gran Paradiso National Park sits right on the border with France, where it constitutes a transfrontier international park together with the Vanoise National Park. Transfrontier protected areas are being formed in a number of countries in Europe, most notably in Eastern Europe where areas in Hungary, Austria, Serbia, Romania, Bulgaria, Albania, Greece, Macedonia, and Croatia are involved, either in mountain chains or along big river systems.
Erlend Haarberg

PAGES 272-273
Alpine Ibex
Capra ibex
ITALY/GRAN PARADISO NATIONAL PARK, VALLE D'AOSTA

Ibex are easily seen by visitors to the Gran Paradiso National Park. Research shows that tourism income and employment in municipalities with national parks is often more than twice that of the average. Suddenly, through nature tourism, otherwise "useless" wilderness areas are becoming economically valuable.
Erlend Haarberg

Here are 21 things you can do to support biodiversity in Europe:

> Buy more certified organically grown food, ideally from local producers.

> Buy only Forest Stewardship Council–certified wood, furniture and paper products. FSC is a label that guarantees that the product comes from certified, responsible forestry.

> Buy only Marine Stewardship Council–certified fish and seafood products. MSC is a label that guarantees that the product comes from certified, responsible fishery.

> Buy only environmentally friendly soaps and washing detergents. Avoid chlorine, chemical bleach, and phosphates.

> Enjoy nature travel, bring your kids along, and pay a fair price for local services.

> Put out food for the birds.

> Put up nest boxes and nesting platforms for birds and bats.

> Create a butterfly garden sown with your favourite wildflowers.

> Plant local species, and get rid of the alien ones.

> Spread seeds from endangered wildflowers along the roadsides and in meadows from which they have disappeared.

> Use no poisons or chemical fertilizers in your garden or on your farm.

> Spend less fossil fuel and create less CO_2 emissions by having your nature holiday nearer to home.

> Plant native trees or donate to forest restoration projects.

> Travel more by train—less by car and airplane.

> Invest in energy-saving in your household.

> Vote for political parties that act in favour of nature and wildlife conservation.

> Choose wine from glass bottles sealed with natural cork stoppers rather than plastic or screw-top stoppers.

> Buy no more medicine than you need, and return unused surplus back to the pharmacy. Don't flush them in the toilet—antibiotics and hormones are very harmful to nature.

> Reuse and recycle more, and use less paper.

> Volunteer to help with research, restoration, and communication projects.

> Donate to biodiversity-conservation projects.

PAGES 274-275
Laurel forests and canary geranium
Laurus azorica and *Geranium canariensis*
SPAIN/GARAJONAY NATIONAL PARK, LA GOMERA, CANARY ISLANDS

Together with La Palma, La Gomera is the most humid of the Canary Islands and is home to a unique forest of laurel species that exist here, on Madeira, and on the Azores. Here the fog hangs on a daily basis, and rains are frequent, creating a subtropical rainforest, a relic of the forests that once covered large parts of the Mediterranean basin 10,000 years ago.

Iñaki Relanzón

LEFT
European wildcat
Felis silvestris
MOLDOVA/CODRII FOREST RESERVE

It is believed that most domestic cat races originated from the African wildcat, but here in Europe, we have our own wildcat species. A native of most of southern, central, and eastern Europe, the wildcat is now endangered in several countries. In Scotland it is estimated that only 400 remain, making it even rarer than the Indian tiger.

Laurent Geslin

RIGHT
Fox and car
Vulpes vulpes
UNITED KINGDOM/LONDON

The red fox is probably Europe's most adaptable and versatile predator, living in virtually all habitats and in every corner of the continent. A great opportunist, it eats almost anything it can find, from garbage, carrion, and rats, to roe deer and red deer fawns. The red fox is quite happy living alongside us and often prospers right in the centre of our major cities, as here in London. Many species can adapt to an urban environment—if we just let them.

Laurent Geslin

PAGES 278-279
European crane
Grus grus
SWEDEN/LAKE HORNBORGA, VÄSTERGÖTLAND

The crane is another comeback success species, with its migration stopover sites quickly becoming local tourism magnets. Here at Lake Hornborga in Sweden, where each spring some 15,000 cranes gather, around 150,000 people come to watch them. The biggest gatherings of cranes in Europe are at Hortobágy in Hungary, outside Berlin and on the Baltic Coast in Germany, and in Montier-en-Der in France. In Hortobágy in 2009, the cranes numbered over 100,000.

Stefano Unterthiner

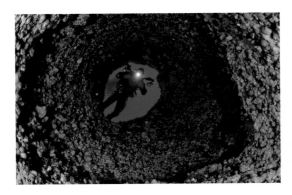

Straight from the Dream Team in the field

We wanted to share with you a few behind-the-scenes images—from the photographers in the field. This incredible team did whatever was necessary to bring you the wonderful images in this book.

On the following page you will find a full list of the Dream Team photographers—the real heroes in this project, without whom it would not have been possible for us to share these Wild Wonders of Europe with you. The photographers were carefully selected for their visual artistry and their commitment to working together to produce something that has never been seen or done before.

LEFT TO RIGHT, TOP TO BOTTOM

Grzegorz Leśniewski with a marmot in Hohe Tauern NP, Austria.

Milán Radisics submerged in Lake Skadar, Montenegro.

Grey seal trying out Laurent Geslin's camera gear, United Kingdom.

Brown bear cub checking out Staffan Widstrand's remote-controlled camera, Finland.

Ole Jørgen Liodden in a −25°C (−13°F) blizzard on Svalbard, Norway.

Christian Ziegler in a close encounter with a chameleon, Greece.

Magnus Lundgren diving in Saltstraumen, Norway.

Orsolya Haarberg among the basalt cliffs, Litlanesfoss, Iceland.

A capercaillie cock attacking photographer Pete Cairns's equipment.

AUSTRIA

Daniel Zupanc
www.zupanc.at

Georg Popp
www.popphackner.com

Verena Popp-Hackner
www.popphackner.com

DENMARK

Jesper Tønning
www.jespertonning.dk

ESTONIA

Sven Začek
www.zacekfoto.ee

FINLAND

Jari Peltomäki
www.finnature.fi

Lassi Rautiainen
www.articmedia.fi

Markus Varesvuo
www.birdphoto.fi

FRANCE

Laurent Geslin
www.laurent-geslin.com

Olivier Grunewald

Vincent Munier
www.vincentmunier.com

GERMANY

Christian Ziegler
www.naturphoto.de

Claudia Müller
www.flowerpics.de

Cornelia Dörr
www.doerr-naturbilder.de

Dieter Damschen
www.dieterdamschen.de

Dietmar Nill
www.dietmar-nill.de

Florian Möllers
www.florianmoellers.com

Frank Krahmer
www.frankkrahmer.de

Ingo Arndt
www.ingoarndt.com

Konrad Wothe
www.konrad-wothe.de

Sandra Bartocha
www.bartocha-photography.com

Solvin Zankl
www.solvinzankl.com

Theo Allofs
www.theoallofs.com

HUNGARY

Bence Máté
www.matebence.hu

László Novák
www.novaklaszlo.hu

Milán Radisics
www.milan.hu

Orsolya Haarberg
www.haarbergphoto.com

ICELAND

Daniel Bergmann
www.danielbergmann.com

ITALY

Bruno D'Amicis
www.brunodamicis.com

Elio della Ferrera
www.eliodellaferrera.com

Manuel Presti
www.wildlifephoto-presti.com

Maurizio Biancarelli
www.mauriziobiancarelli.net

Stefano Unterthiner
www.stefanounterthiner.com

NETHERLANDS

Edwin Giesbers
www.edwingiesbers.com

Ruben Smit
www.rubensmit.nl

NORWAY

Erlend Haarberg
www.haarbergphoto.com

Kai Jensen
www.villmarksbilder.com

Nils Aukan
www.aukanphoto.no

Ole Jørgen Liodden
www.naturfokus.no

Pål Hermansen
www.palhermansen.com

Tom Schandy
www.tomschandy.no

POLAND

Grzegorz Leśniewski
www.grzegorzlesniewski.pl

PORTUGAL

Luís Quinta
www.luisquinta.net

Nuno Sá
www.photonunosa.com

RUSSIA

Igor Shpilenok
www.shpilenok.com

Sergey Gorshkov
www.gorshkov-photo.com

SLOVENIA

Arne Hodalič
www.arnehodalic.si

SPAIN

Diego Lopez
www.gonature.es

Iñaki Relanzón
www.photosfera.com

José B. Ruiz
www.josebruiz.com

Juan Carlos Muñoz
www.artenatural.com

SWEDEN

Anders Geidemark
www.andersgeidemark.se

Magnus Elander
www.magnuselander.com

Magnus Lundgren
www.aquagraphics.se

Martin Falklind
www.fiskejournalen.se

Mireille de la Lez
www.novaphotomedia.com

Peter Lilja
www.peterlilja.com

Staffan Widstrand
www.staffanwidstrand.se

SWITZERLAND

Franco Banfi
www.banfi.ch

Michel Roggo
www.roggo.ch

UNITED KINGDOM

Danny Green
www.dannygreenphotography.com

David Maitland
www.davidmaitland.com

Laurie Campbell
www.lauriecampbell.com

Linda Pitkin
www.lindapitkin.net

Mark Carwardine
www.markcarwardine.com

Mark Hamblin
www.markhamblin.com

Niall Benvie
www.imagesfromtheedge.com

Pete Oxford
www.peteoxford.com

Peter Cairns
www.northshots.com

WILD WONDERS OF EUROPE

Unseen, unexpected, unforgettable
Revealing Europe's amazing natural treasures to the world.
A heritage to share. To enjoy. To protect.

www.wild-wonders.com

Dream Team Partners

These are the Dream Partners that made
our "Mission Impossible" possible.

MAIN PARTNERS
WWF www.panda.org
National Geographic Society www.nationalgeographic.com
Epson www.epson.com
Nokia www.europe.nokia.com/home
Nikon www.europe-nikon.com

GOLD PARTNERS
Bayard Presse www.groupebayard.com/fr

SILVER PARTNERS
Nature Picture Library (NPL) www.naturepl.com
Conservation International (CI) www.conservation.org
Eazyfone www.eazyfone.com
SanDisk www.sandisk.com
Gitzo www.gitzo.com
Life Exhibitions www.lifeexhibitions.com
PAS Events www.pasevents.com

LOCAL PARTNERS
PolarQuest www.polar-quest.com Aunan Lodge www.aunan.no Norway
Nature www.norway-nature.com Articmedia www.articmedia.fi Wild Taiga
www.wildtaiga.fi Finnature www.finnature.fi Noble Caledonia
www.noble-caledonia.co.uk Flatanger Kommune www.flatanger.kommune.no
Innovation Norway www.innovasjonnorge.no Kuusamo Lapland
www.kuusamolapland.fi Forollhogna www.forollhogna.org Äventyrsresor
www.aventyrsresor.se VisitFinland www.visitfinland.com Naturfokus
www.naturfokus.no Destinasjon Saltstraumen www.destinasjon-saltstraumen.com
Reisen in die Natur www.reisen-in-die-natur.de Objectif Nature
www.objectif-nature.com Lanzarote www.turismolanzarote.com VisitSweden
www.visitsweden.com Andorra Tourisme www.andorra.ad Nordic Discovery
www.nordicdiscovery.se Parque Natural da Madeira www.pnm.pt Madeira
Islands www.madeiraislands.travel Ulvsbomuren www.ulvsbomuren.se King of
the Forest www.skogenskonung.se Staatsbosbeheer www.staatsbosbeheer.nl
Azores Tourism www.azorestourism.com Mediwiss Analytic GmbH
www.mediwiss-analytic.de Islas Canarias www.turismodecanarias.com Nordic
Safari www.nordicsafari.net Valle d'Aosta www.regione.vda.it Swedish
Environmental Protection Agency www.naturvardsverket.se Visit Malta
www.visitmalta.com Zermatt Tourismus www.zermatt.ch Padi Dive Center
Iceland www.dive.is Slovenian Tourist Board www.slovenia.info

ENDORSING PARTNERS
The World Conservation Union (IUCN) www.iucn.org European
Environment Agency (EEA) www.eea.europa.eu International League of
Conservation Photographers (ILCP) www.ilcp.com United Nations
Environment Programme (UNEP) www.unep.org Convention on Biological
Diversity (CBD) www.cbd.int The WILD Foundation www.wild.org PEBLDS
www.peblds.org BirdLife International www.birdlife.org Countdown 2010
www.countdown2010.net EuroNatur www.euronatur.org AEFONA www.aefona.org
PAN Parks Foundation www.panparks.org The Alpine Convention www.alpconv.org
EAZA European Carnivore Campaign www.carnivorecampaign.eu
Fotonatura.org www.fotonatura.org Visión Salvaje www.euromodelismo.com
Image et Nature www.image-nature.com Naturapics www.naturapics.com
NVN www.nvnfoto.nl Fotonaturis www.fotonaturis.org Grasduinen
www.grasduinen.nl IFWP www.ifwp.net Birding in Spain www.birdinginspain.com
Photo Travel Romania www.phototravel.ro Hidephotography.com
www.hidephotography.com Fotodelta www.fotodelta.ro Sakertour www.sakertour.com
Salix Nature Tours Ukraine www.salix.od.ua EkoClub www.ekoclub.com
Neophron Tours www.neophron.com Asferico www.asferico.com Natuurgek
www.natuurgek.nl Light & Nature www.lightandnature.hu NaturArt www.naturart.org

RIGHT
Cyprus tulip
Tulipa cypria
CYPRUS/AKAMAS PENINSULA
Another one of the original wild tulips,
found on Cyprus and nowhere else.
Peter Lilja

Thank you all!

First of all, this book is dedicated to the many generations of nature conservationists, from Queen Eleonora of Sardinia way back in 1395, and onwards, for your devoted struggle, tough negotiations, and public-awareness-raising. Thanks to you all, we still have a shared European natural heritage to enjoy.

To be able to create this, the biggest nature photography project in history, we teamed up with a large number of devoted, knowledgeable, and positive-minded institutions, organisations, ministries, cities, EU offices, corporations, and individuals. We truly owe all of you our profound gratitude which we would like to express here. Not through mentioning each and every single person that really deserves it, in big and small ways, since that list would be many pages long, but through humbly thanking you here, all at the same time.

You were all part of making our Mission-Impossible project possible.

We wouldn't have gotten anywhere without you.

Having said that, some people should indeed be further mentioned:

Our Dream Team of talented, hard-working, and loyal photographers, without whom none of the pictures in this book would have been taken.

And our Dream Team of loyal and committed partners, who by chipping in to help us out in a multitude of different ways, made it possible to send the photographers out on their assignments and to create something out of what they brought back. Among them, we would especially like to thank our main partners—the WWF International, the National Geographic Society, Nokia, Epson, and Nikon. You have been the backbone of our project-partner team and we are very happy for your cooperation.

And we would like to thank the WILD Foundation and the ILCP, who during the World Wilderness Congress gave us direct inspiration to start creating Wild Wonders of Europe.

We would also like to mention those who have done their invaluable part for us as staff, consultants, or advisors: Lena Sundqvist, Karin Wehlin, Daniel Ahlbert, Magnus Sylvén, Neil Wakeling, Danny Sweeney, Roz Kidman-Cox, Niall Benvie, Karolina Eljas, Hasse Berglund, Fredrik Hyltén-Cavallius, Emma Blyth, Dorothea Erpenbeck, Anton Spetz, Marc Hesse, Erika Tirén, Erik Collinder, Andy Langley, and Annika Backman.

Warm thanks also to Anne-Marie Bourgeois who is the genius behind the beautiful design of this book, and to the editorial team at Éditions de La Martinière, who took this book under their efficient wings and are spreading it over the world.

Finally, we want to convey a deeply felt thanks to our families and loved ones, who during the last years have seen far too little of us, and who have shown remarkable patience, understanding, and support for us while we were carrying out this almost epic undertaking. We love you!

Thank you again, all of you, for helping us to make the Wild Wonders of Europe visible to the world. Our natural heritage will be deeply grateful to you for that endeavour, and for aeons ahead.

The Wild Wonders Map of Europe

Europe is the second-smallest continent on Earth, measuring about 10,180,000 sq. km—a continent whose exact borders are not always clearly defined. We decided for a mainly classic geographical definition, including the Mediterranean north coast and islands, the Black Sea coast, the high Caucasus mountains, the Ural mountains, Greenland, and the Atlantic islands as far away as Madeira, the Azores and the Canary Islands.

Greenland
[Denmark]

Atlantic Ocan

1 Azores
[Portugal]

2 Madeira
[Portugal]

PORTUGAL

3

9

8

10

2

Canary Islands
[Spain]

5 **11**

4

Arctic Ocean

Franz Josef Land

Svalbard
(Norway)

Novaya Zemlya

Greenland Sea

Barents Sea

2
CELAND

1

Norwegian Sea

3

1 **2**
5

3 **2**

4 **6**
1

6

FINLAND

3

RUSSIA

4 **2** **7** **8**

5

SWEDEN

NORWAY

Faroe Islands
(Denmark)

5

6 **4**

4
3

ESTONIA

1

North Sea

5

LATVIA

2

ELAND

UNITED KINGDOM

2

Baltic
Sea

1

LITHUANIA

Kaliningrad
(Russia)

BELARUS

1

3

1

3

DENMARK

1

3 **2**

2

2 **1**
3

1

POLAND

UKRAINE

1

NETHERLANDS
BELGIUM

GERMANY

LUXEMBOURG

CZECH
REPUBLIC

SLOVAKIA

4 **2**
2

GEORGIA

1

FRANCE   LIECHTENSTEIN

SWITZERLAND

3

AUSTRIA

2 **2**

1

HUNGARY

1

1

MOLDOVA

3

6

7

3

2

5 **2**

1

3

1

ANDORRA

12

MONACO

7

SLOVENIA

2

3

CROATIA

BOSNIA AND
HERZEGOVINA

2

1

SERBIA

ROMANIA

2

1

1

BULGARIA

2

Black Sea

TURKEY

SAINT-MARIN

1

2

Kosovo

3

1

SPAIN

6

Corsica

6

ITALY

4

VATICAN CITY

4

MONTENEGRO

1

MACEDONIA

ALBANIE

1

2

GREECE

1

5 Sardinia

1

3

Cypress

3

4

Mediterranean Sea

Sicily

Peloponnesia

Crete

2

2

MALTA

This book is one of the many results of the Wild Wonders of Europe project, on the Web at www.wild-wonders.com.

All photographs were made in the wild, of free and wild beings, except the captive tuna, on pages 8–9 and 212, and the wolves on pages 137 and 140–141. Photographs on pages 10–11 and 93 were made in large, fenced reserves.

All images from this book are available through the Nature Picture Library at www.naturepl.com. Many images are also available as prints through www.wild-wonders.com and from Art for Conservation at www.artforconservation.org. Wild Wonders also has a series of outdoor and indoor exhibitions and AV-shows to offer venues big and small.

PREPRESS: Point4, Paris, France
FACT CHECKER: Dr. Magnus Sylvén
TRANSLATORS: Ariel Marinie and Cécile Dutheil de la Rochère

ENGLISH-LANGUAGE EDITION
PROJECT MANAGER: Aiah Rachel Wieder
DESIGNER: Shawn Dahl
PRODUCTION MANAGER: Jules Thomson

Cataloging-in-Publication Data has been applied for and may be obtained from the Library of Congress.
ISBN: 978-0-8109-9614-4

Printed and bound in France
10 9 8 7 6 5 4 3 2 1

Abrams books are available at special discounts when purchased in quantity for premiums and promotions as well as fundraising or educational use. Special editions can also be created to specification. For details, contact specialmarkets@abramsbooks.com or the address below.

ABRAMS
THE ART OF BOOKS SINCE 1949

115 West 18th Street
New York, NY 10011
www.abramsbooks.com

ARJOWIGGINS
GRAPHIC

PAGES 284–285
White storks nesting on an exposed Atlantic cliff
Ciconia ciconia
PORTUGAL/SOUTHWEST ALENTEJO AND VICENTINO COAST NATURAL PARK

You can't get much farther west in Europe than Portugal's Cabo Sardão. Jutting out into the mighty Atlantic Ocean, exposed to the brutality of the elements, these cliffs are quite surprisingly the preferred nesting site for a colony of white storks. The white stork is one of Europe's most emblematic birds, the bringer of babies in several cultures and connected with good fortune in ancient folklore. It has made a remarkable comeback during the last 15 years, with numbers up by over 300 per cent in Spain. The white stork is also the national bird of Lithuania, Belarus, and Poland.
Luís Quinta

LEFT
Great white pelican
Pelecanus onocrotalus
ROMANIA/DANUBE DELTA BIOSPHERE RESERVE, TULCEA

White pelicans breed from southeastern Europe to Africa and India, with the biggest colonies in the world in the Romanian-Ukrainian Danube delta. At 6,200 sq. km (2,400 sq. mi.), this Unesco World Heritage Area and biosphere reserve is one of Europe's most important wetlands and the second largest delta in Europe, after the Volga delta. Ecotourism is increasing here, with visitor options including boat trips into the vast delta, floating hotels, camps, farm stays, and birdwatching hides.
Manuel Presti

Ancient wildlife art for the chapter pages

PAGE 18
Flying fish
Exocetus volitans
GREECE/PHYLAKOPI, MILOS
C. 2,000 BCE

PAGE 28
Elk or moose
Alces alces
SWEDEN/GLÖSA
C. 3,000 BCE

PAGE 80
Flower
GREECE/CRETE, KNOSSOS
C. 500 BCE

PAGE 128
Aurochs bull
Bos primigenius
FRANCE/LASCAUX
C. 13,000 BCE

PAGE 186
Raven
Corvus corax
UNITED KINGDOM/SCOTLAND
C. 1100 CE

PAGE 246
Red deer hunting
Cervus elaphus
FRANCE/LASCAUX
C. 13,000 BCE